Remington Education

Drug Information and Literature Evaluation

Remington Education

The *Remington Education* series is a new series of indispensable guides created specifically for pharmacy students. Providing a simple and concise overview of the subject matter, the guides aim to complement major textbooks used.

The guides assist students with integrating the science of pharmacy into practice by providing a summary of key information in the relevant subject area along with cases and questions and answers for self assessment (as subject matter allows). Students will be given a practical way to check their knowledge and track the progress of their learning before, during, and after a course.

Key features of the texts include:

- Learning Objectives
- Tips and Tricks Boxes
- Key Points Boxes
- Illustrations
- Assessment Questions
- Further Reading

The *Remington Education* series is a must-have for all pharmacy students wanting to test their knowledge, gain a concise overview of key subject areas and prepare for examinations.

Covering all areas of the undergraduate pharmacy degree, the first titles in the series include:

- *Drug Information and Literature Evaluation*
- *Introduction to Pharmacotherapy*
- *Physical Pharmacy*
- *Law & Ethics in Pharmacy Practice*
- *Pharmaceutics*

Remington Education
Drug Information and Literature Evaluation

Marie A. Abate BS, PharmD
Professor of Clinical Pharmacy, Director, WV Center for Drug and Health Information, Director for Programmatic Assessment, West Virginia University School of Pharmacy, Morgantown, WV, USA

Matthew L. Blommel PharmD
Clinical Associate Professor, Assistant Director, WV Center for Drug and Health Information, West Virginia University School of Pharmacy, Morgantown, WV, USA

Published by Pharmaceutical Press

1 Lambeth High Street, London SE1 7JN, UK

© Royal Pharmaceutical Society of Great Britain 2013

 is a trade mark of Pharmaceutical Press

Pharmaceutical Press is the publishing division of the Royal Pharmaceutical Society

Typeset by Laserwords Private Limited, Chennai, India
Printed in Great Britain by TJ International, Padstow, UK
Index provided by Indexing Specialists, Hove, UK

ISBN 978 0 85711 066 4

All rights reserved. No part of this publication may be reproduced, stored in a retrieval system, or transmitted in any form or by any means, without the prior written permission of the copyright holder.

The publisher makes no representation, express or implied, with regard to the accuracy of the information contained in this book and cannot accept any legal responsibility or liability for any errors or omissions that may be made.

The rights of Marie A. Abate and Matthew L. Blommel to be identified as the authors of this work have been asserted by them in accordance with the Copyright, Designs and Patents Act, 1988.

A catalogue record for this book is available from the British Library.

Library of Congress Cataloging-in-Publication Data has been requested

Contents

Preface		ix
About the authors		xi
Introduction		xiii

1 Evidence-based approach to addressing information needs — 1
  Introduction — 1
  A seven-step systematic approach — 2
  Summary — 11

2 Tertiary resources — 13
  Introduction — 13
  General drug information resources — 14
  Information-specific resources — 18
  Summary — 27

3 Secondary resources — 29
  Introduction — 29
  Searching secondary resources — 30
  Commonly used secondary resources — 33
  Summary — 37

4 Internet resources — 39
  Introduction — 39
  Evaluating the quality of internet-based medication and health information — 40
  Internet resources — 44
  Useful internet sites for clinicians — 45
  Summary — 52

5 Evaluating clinical studies – step 1: what type of study is it? — 53
  Introduction — 53
  Experimental studies — 53
  Observational studies — 54
  Summary — 60

6 Evaluating clinical studies – step 2: the journal, authors, and study purpose — 61
Introduction — 61
Journal and investigator considerations — 61
Potential conflicts of interest — 62
Study objectives and hypotheses — 64
Summary — 67

7 Evaluating clinical studies – step 3: methods used — 69
Introduction — 69
Eligibility (inclusion and exclusion) criteria — 69
Sampling (enrollment) considerations — 71
Informed consent — 72
Sample size — 73
Controlled experimental designs — 74
Assignment to interventions — 77
Blinding — 77
Treatment considerations — 80
Outcomes and variables — 83
Measurements — 84
Summary — 90

8 Evaluating clinical studies – step 4: statistical analyses — 91
Introduction — 91
One-tailed versus two-tailed tests — 92
Categories of statistical tests — 92
Parametric statistical tests — 93
Nonparametric statistical tests — 95
Correlation — 98
Regression — 100
Summary — 103

9 Evaluating clinical studies – step 5: results, interpretation, and conclusions — 105
Introduction — 105
Measures of central tendency — 106
Measures of variability (spread or dispersion) — 108
Statistical inference — 111
Confidence interval — 112
Hypothesis testing — 114
P values — 115
Type I error — 117
Type II error — 118
Statistical significance versus clinical significance (importance) — 121

	Measures of risk, risk reduction, and clinical utility	122
	Drop-outs and data handling	128
	Discussion section	131
	Summary	136
10	**Evaluating clinical studies – step 6: putting it all together**	**137**
	Introduction	137
	Key questions and considerations when critiquing published experimental studies	137
11	**Equivalence and noninferiority studies**	**143**
	Overview	143
	Equivalence study	144
	Noninferiority study	145
	Summary	150
12	**Practice guidelines, systematic reviews, and meta-analyses**	**151**
	Introduction	151
	Systematic reviews	151
	Meta-analyses	152
	Practice guidelines	155
	Summary	159

Answers to self-assessment questions	161
Bibliography	183
Index	185

Preface

This book is designed for healthcare professionals to serve as a guide to using and evaluating drug information resources in practice. It provides an overview of commonly used information sources available in print, online, or in mobile versions, with their advantages and limitations, as well as database resources (e.g., PubMed, IPA) used to locate primary literature (clinical studies).

An overview of types of clinical studies and systematic reviews is included, with several chapters devoted to providing key points and concise, practical considerations for critically analyzing and evaluating all aspects of clinical drug studies. Each book chapter includes worked examples and questions with answers to assist with learning and applying important concepts.

Marie A. Abate
Matthew L. Blommel
2012

About the authors

Dr. **Marie Abate** graduated from the University of Michigan College of Pharmacy. She is nationally known in the area of drug information and has served as Director of the statewide West Virginia Center for Drug and Health Information (WV CDHI) for the past 20 years. As part of a grant received from the US Department of Education's Fund for the Improvement of Postsecondary Education, Dr. Abate and colleagues developed an innovative computer educational program to teach students how to interpret and assess medical literature, which has been in use at West Virginia University (WVU) for over 15 years and has also been used by other schools of pharmacy. Dr. Abate has authored several publications in the areas of education and drug information, and she has also served as Director of Assessment for the school since 2004.

Dr. **Matthew Blommel** graduated from Mercer University Southern School of Pharmacy and completed a Pharmacy Practice Residency at the University of Pittsburgh Medical Center. He has served as Assistant Director of the statewide West Virginia Center for Drug and Health Information (WV CDHI) for the past 8 years. During this tenure, he has precepted more than 150 students in an Advanced Practice Pharmacy Experience Drug Information Rotation. He also teaches an Introduction to Pharmacy Resources session, as well as in-depth lectures on various pharmacy-related information resources in the school's didactic curriculum.

Introduction

This book serves as a concise guide to medical literature evaluation and the provision of medication and health information by healthcare professionals. The topics covered include the appropriate way to approach a question, the various types of available tertiary and secondary information resources, and important considerations for evaluating and analyzing clinical studies. This book is not intended to train medication information specialists, but rather to allow practitioners and students to gain insight into the process of properly responding to information requests, evaluating information resources, and critiquing studies that can be used in any area of practice. One does not need to be a medical/drug information specialist to perform the skills and functions described in this book since they are integral to every practitioner despite the practice setting.

A brief description follows of the content within each of the chapters in this book. In general, the book begins with chapters that describe the process of responding to medication information requests and the commonly used resources for locating relevant information. The later chapters focus on the evaluation of clinical studies that would be found in the primary literature (e.g., journals).

Chapters

Chapter 1: Evidence-based approach to addressing information needs

Chapter 1 summarizes the importance of using an evidence-based approach when responding to medication and health information questions. It also lists the steps of a sample modified systematic approach, and describes the function and purpose of each of the steps listed.

Chapter 2: Tertiary resources

Chapter 2 discusses the use of tertiary resources to locate information. It provides a description of various tertiary resources, organized by the type of information each tertiary resource is focused on.

Chapter 3: Secondary resources

Chapter 3 gives an overview of secondary resources of information. It discusses general methods of using the secondary resources to locate primary literature, including controlled vocabularies, combining search terms, and identifying the key terms within the question being researched. This chapter also describes the various secondary resources commonly used.

Chapter 4: Internet resources

Chapter 4 provides a discussion on the use of internet-based sources of medical and health information and the limitations of the material found. It describes the characteristics that should be considered when evaluating internet-based material and provides some websites that may be especially useful to clinicians.

Chapter 5: Evaluating clinical studies – step 1: what type of study is it?

Chapter 5 provides an overview of the main types of study designs, focusing on the observational designs and their features, advantages, and disadvantages.

Chapter 6: Evaluating clinical studies – step 2: the journal, authors, and study purpose

Chapter 6 focuses on methods for helping to ensure journal quality, potential conflicts of interest for a study's authors/investigators, and the introduction of a published study, including hypotheses and the study objective.

Chapter 7: Evaluating clinical studies – step 3: methods used

Chapter 7 discusses study eligibility criteria, informed consent, types of controlled experimental designs, assignment to study groups, adherence, and measurement considerations. It also introduces types of variables and scales of measurement for study data.

Chapter 8: Evaluating clinical studies – step 4: statistical analyses

Chapter 8 provides an overview of statistical tests, included one-tailed versus two-tailed tests, parametric versus nonparametric tests, and correlation and regression analyses. The focus is on understanding the purpose and appropriate use of these tests and not on their actual calculation.

Chapter 9: Evaluating clinical studies – step 5: results, interpretation, and conclusions

Chapter 9 focuses on a study's data, including measures of central tendency (means, medians, mode) and measures of variability. Statistical inference is discussed, including confidence intervals, power, and possible errors when making conclusions. Measures of risk and clinical utility are addressed, along with patient drop-outs, data-handling methods with drop-outs, and a study's discussion and conclusions sections.

Chapter 10: Evaluating clinical studies – step 6: putting it all together

Chapter 10 serves as a summary of the important concepts addressed in Chapters 7–9 by providing bulleted lists of key questions and considerations when evaluating a published experimental study.

Chapter 11: Equivalence and noninferiority studies

Chapter 11 provides a brief overview of equivalence and noninferiority studies, which differ from the more commonly reported "superiority" experimental studies in the literature.

Chapter 12: Practice guidelines, systematic reviews, and meta-analyses

Chapter 12 discusses other types of publications of value to clinicians, including practice guidelines, systematic reviews and meta-analyses, how they are prepared, potential problems with their use, and readily available sources to use for locating these publications.

Brief overview of medication information

The first drug information center was established in the early 1960s. Since that time, numerous other drug information centers have been developed. The clinicians who practice in these centers are known as drug information specialists and usually have training or experience specifically in the practice of providing medication-related information through a variety of means. However, it is important that healthcare practitioners in a variety of fields, including pharmacy, medicine, dentistry, and other areas, possess the knowledge and skills to locate, access, review, evaluate, interpret, and apply drug or health-related information.

It was through the early provision of drug information that pharmacists in particular began to demonstrate the value of their ability to serve as information resources and to use that information to help guide therapeutic decisions as part of a healthcare team. Thus, drug information provision was perhaps the impetus and basis for developing the clinical pharmacist specialist positions that are an integral part of many healthcare settings today. In addition, community-based pharmacists have been providing information to healthcare providers and patients for decades. This information was initially focused on drug product compounding and dispensing. However, many community pharmacists have become an integral part of a larger healthcare team to improve patient outcomes by establishing broad-based, patient-focused care programs in conjunction with collaborating physicians. In each setting, the need for drug information has expanded along with pharmacists' roles. It is paramount for pharmacists in diverse practice settings to be proficient not only in providing information, but also in having the ability to analyze and evaluate published literature and to develop recommendations based on the best available data.

The term medication information, or medication and health information, is commonly used in current practice settings instead of drug information. This change in terminology was made primarily to reflect the broader role that all pharmacists play in providing medication information, as well as to indicate the importance of information use in properly managing medication therapy. Appropriate information use is in fact a cornerstone of the practice of evidence-based medicine across health disciplines when providing patient care.

There are various other responsibilities of clinicians on the healthcare team for which a working knowledge of medication information and medical literature evaluation concepts is integral. For example, adverse drug event monitoring and reporting involve thorough knowledge of potential adverse events and evaluation of any actual events that occur. This evaluation generally includes reviewing patient-specific information, available literature discussing the adverse event, and the use of clinical judgment and a scale such as the Naranjo scale to assess the probability of the adverse event. The practice of medicine and pharmacy involves lifelong learning about advances in pharmacotherapeutics that necessitates ongoing use of information resources. Pharmacy and therapeutics (P & T) committees establish and maintain medication formularies and make decisions concerning rational drug use within healthcare institutions. Responsibilities of a P & T committee include the preparation of monographs for medications being reviewed, development of drug therapy guidelines, review of medication safety data, and conduct of medication use evaluations. The access and appropriate use of information are critical to all of these responsibilities. An understanding of research study designs is obviously necessary for clinician participation in clinical and practice-based research trials.

In summary, the ability to apply in practice the concepts and skills discussed in this book is not limited to those clinicians who are designated "medication information specialists." Rather, all healthcare professionals should be able to locate and evaluate comprehensive, up-to-date information on which to base patient care and other important decisions.

1

Evidence-based approach to addressing information needs

Learning objectives

Upon completion of this chapter, you should be able to:

- discuss the importance of using a systematic, evidence-based approach to answering drug and health information questions
- identify the steps of a systematic approach to answering drug and health information questions
- understand the importance of obtaining background information for questions
- construct an appropriate search strategy to find relevant information
- discuss limitations of tertiary and secondary resources
- formulate and communicate a logical and rational response to a medication and health information question.

Introduction

Health professionals frequently have medication- or therapy-related questions that arise during practice. However, clinicians often seek simple, short answers to their questions due to time constraints, the impression that informal questions or information needs do not require in-depth research and analysis, or a lack of training in locating medication information in a systematic manner. Many simple questions turn out to have more complex answers if an appropriate, thorough approach to finding relevant information is utilized. The process used to answer a medication- or health-related question should be the same regardless of the context in which the question arose or was received, the perceived simplicity of the question or need, or the individual asking the question.

Evidence-based medicine refers to the use of current best evidence in clinical decision making, which requires the incorporation of the strongest external evidence with clinical expertise. Key components of evidence-based medicine include critical literature appraisal, clinical judgment, knowledge about the involved patient, and the use of a sound scientific basis for decision making when such evidence exists. The steps involved in the provision of evidence-based medicine include:

1. forming specific clinical questions
2. searching for and retrieving the "best" evidence
3. critically appraising the evidence for strength, validity, usefulness, and importance
4. summarizing and applying the results to clinical practice
5. evaluating performance and follow-up for improvement.

While the term "evidence-based medicine" is fairly contemporary, a systematic approach to answering drug information questions was first introduced in 1975 to aid in the training of students and pharmacists not adept at providing medication information. Since the introduction of this systematic approach, several modifications have been made to elaborate on the initial concepts and to elucidate needs further when answering information requests.

The number of steps involved in the different versions of the systematic approaches to answering questions usually ranges from seven to nine. Some versions combine similar concepts to produce fewer steps, while others separate out certain aspects into more steps. Regardless of the number of steps, all of the versions involve the same basic concepts that are analogous to the provision of evidence-based medicine in clinical practice.

Key Point

There are several modified systematic approaches to handling information questions that employ evidence-based methods. The exact version used is not as important as the fact that each involves an organized process. Select an approach based on personal preference, professional expectations, and what works best for the involved practice setting.

This chapter focuses on the use of one of the modified systematic approaches to answering medication-related questions, to ensure that these types of questions are handled consistently regardless of the clinician locating the information, the specific question/information need, or the context in which a question is asked. The specific steps described here for answering medication-related questions mimic those steps involved in the use of evidence-based medicine to resolve clinical problems. The steps should not be treated as individual components, but instead should progress fluidly and in a manner that allows interaction with a requester to be efficiently maintained.

A seven-step systematic approach

Table 1.1 outlines a seven-step systematic approach to respond to medication information questions or needs. Each step is detailed in the following sections.

Table 1.1 A modified systematic approach for addressing medication-related questions

Step 1	Obtain requester demographics
Step 2	Obtain pertinent background information
Step 3	Classify the ultimate question
Step 4	Systematic search for information
Step 5	Evaluate, analyze, and synthesize information found
Step 6	Provide response
Step 7	Follow-up and document

Obtain requester demographics

> **Key Point**
>
> Other important information to obtain when one receives an information request includes:
>
> - how best to communicate with the requester (e.g., phone number, fax number, e-mail address)
> - an assessment of the urgency of the request, and the desired delivery method of the response (verbal or written)
> - the requester's affiliation or practice site (if a healthcare provider is the requester)
> - when the request was made
> - the requester's location (if not at the same site as the clinician who is researching the answer to the request).

When responding to an information request, the first step should be to gather general demographic information about the requester. This is important because it allows you to assess immediately the level of understanding of the requester. Most medication-related questions received by pharmacists are from either other healthcare providers or the lay public. Responses to questions should be directly related to the level of understanding of the requester; thus, an answer for a drug interaction question may look markedly different when provided to a healthcare provider than when provided to a patient. Even when the requester is initially categorized as healthcare provider or lay public, further questioning can be important to assess the requester's background knowledge and familiarity with the topic. Healthcare professionals, even those practicing in the same discipline, can vary greatly in their experiences and training. Assessing the sophistication of the requester allows information to be provided that best meets an individual's needs.

Obtain pertinent background information

> **Key Point**
>
> If an information request pertains to a specific patient, the following patient-specific background information might be required in order to formulate an appropriate response:
>
> - age
> - gender
> - weight
> - relevant laboratory data (to assess renal and hepatic function, appropriateness of any drug levels obtained, etc.)
> - specific medical diagnosis
> - relevant past medical history, family and social history
> - medications (current or recent, nonprescription, prescription, natural products)
> - allergies.

Obtaining background information should be part of a seamless transition from gathering the requester's demographic information to a focus on the question itself. This is often the hardest step for students and clinicians unfamiliar with providing or locating medication- or therapy-related information on a regular basis. The responder must utilize good communication skills and ask thoughtful, pertinent questions of the requester to characterize fully the precise information need. Identifying the type of background information required and, hence, the type of background-related questions to ask, largely depends on the specific question being asked (or the desired information needed) and the clinical experience of the clinician responder. Without a broad base of experience, it may be difficult for a clinician to know precisely what background information would be pertinent to a question. In general, clinicians should consider why a certain question has arisen and how the requester is planning to use the response as guidelines to formulating background-related questions to ask. When a clinician receives a request for medication or therapy information, the first question that should always be addressed is whether a specific patient is involved or whether the request is general or academic in nature.

Inquiring about the resources a requester has already consulted to locate information is another general background question that can be useful to help reduce duplication of efforts. However, it should not be assumed that just because an individual has already checked particular resources, that those resources were used effectively. Depending on the requester's level of information "sophistication," the responder may need to double-check resources already said to be used.

Other background questions that are relevant to an inquiry may be more difficult for the responder to formulate, particularly when the inquiry involves a topic for which the responder has limited familiarity. Fortunately, the ability to refine inquiries by gathering specific, pertinent background information usually improves with practice and experience.

> **Key Point**
>
> After all pertinent background information has been obtained and the actual question and information need have been identified, it is important for the responder to restate the question to the requester. This allows for the appropriate background information to be incorporated into the answer and to ensure that all involved have a clear understanding of the information need and the actual question being asked.

Ultimately the goal of the background information step is to allow the responder to obtain any additional data needed to answer the question and to determine the "real" question that needs to be addressed. Many initial information requests are clear and represent the actual information need. Other times, however, the initial request does not really state the exact information that is required. In these situations, it is only through appropriate dialogue and obtaining pertinent background information that the actual need is revealed. For this reason, obtaining background information is paramount for providing a thorough, efficient, and appropriate response to address a question.

> **Key Point**
>
> Even when requesters do not volunteer additional information, it is the responder's responsibility to ask for the additional background information necessary to respond properly. Do not assume that a question does not relate to a specific patient. That assumption often leads to an incomplete response that does not truly fulfill the actual information need, and it may lead to misinformation that could adversely affect patient care.

A potential problem when attempting to obtain background information can arise when dealing with a requester who is not the ultimate individual in need of the information. An example of this situation is when a physician asks a pharmacist a question, and the pharmacist in turn contacts a drug information specialist for assistance. Oftentimes this "third-party" or "intermediary" practitioner does not have all of the information needed to answer the specialist's background-related questions. As a result, the intermediary will have to go back to the original requester for the information needed, or the specialist will need to speak to the original requester directly to obtain the information needed.

Worked example

Example 1.1

Suppose a dentist contacts a pharmacist and asks if carvedilol is associated with the development of gingival hyperplasia. The pharmacist in question is busy with other responsibilities but pauses for a moment to look up carvedilol in a tertiary resource that is on hand. The pharmacist does not see gingival hyperplasia listed as an adverse effect for carvedilol and provides this information to the dentist. The dentist accepts the response, and the pharmacist goes back to his duties. Is there a potential problem here?

While the pharmacist technically provided correct information in response to the question asked, there is no way of knowing if the answer met the dentist's actual information need due to the manner in which the pharmacist approached the information request. If the pharmacist had attempted to obtain pertinent background information, the above encounter could have gone much differently.

The following dialogue represents important background questions, with answers provided, that would have helped to clarify the problem appropriately:

Have you looked in any resources to find information for this question?
Answer: I have not used any other resources.

Is this regarding a specific patient?
Answer: Yes.

How long has the patient had the gingival hyperplasia?
Answer: It is unknown exactly. This is the first time the patient has been to my office since his last check-up 1 year ago when he did not have signs of gingival hyperplasia.

How long has the patient been taking carvedilol?
Answer: Six months.

What other medications is the patient taking?
Answer: Fortamet and Exforge.

How long has the patient been taking those medications?
Answer: Three years for the Fortamet and over a year for the Exforge.

What is the patient's age?
Answer: 54 years old.

What is the patient's brushing/flossing habit?
Answer: The patient states he brushes once a day usually, sometimes twice daily. He rarely flosses.

Has the patient ever had this problem in the past?
Answer: No.

The above dialogue was not overly time-consuming, but it did elaborate on the information given in the original question to determine fully the background for the situation and the information needed by the requester. This additional information would have allowed the pharmacist to check all of the patient's medications for the potential adverse effect of gingival hyperplasia. Upon review of the other medications, the pharmacist would have discovered that the amlodipine component of Exforge has been linked to gingival hyperplasia and could be the cause of this patient's condition. This additional information would also have allowed the dentist to address the patient's problem properly. Thus, the original answer provided by the pharmacist did not address the dentist's true information need and, more importantly, it did not benefit the patient.

Classify the ultimate question

After the actual question or need has been identified from the pertinent background information, it can then be classified. Classifying the question requires a thoughtful determination of its area(s) of focus. A common method of classifying medication and health information questions is by type, such as adverse reaction, therapeutic use, pharmacokinetics, dosing, drug interaction, product identification or availability. Properly classifying the question by type mainly serves as a guide to the appropriate resources to use when researching the question, thus allowing for a more efficient search process. For example, if the question or information need relates to compounding a formulation, a clinician should focus on using compounding specific resources when researching the answer. See Chapter 2 for a discussion of the various types of resources used for specific types of questions or information needs. Another benefit of classifying questions in this manner is that it allows one to determine the types of questions that arise most often in a practice. This information could be useful when identifying the types of resources to have available in the future.

Systematic search for information

> **Key Point**
>
> Consider the following when evaluating the validity of information found in tertiary resources:
>
> - At least two (if not more) reputable tertiary resources should be consulted. Studies have shown discrepancies and variability in the information contained in commonly used tertiary resources (Vidal *et al.* 2005; Vitry 2007). Thus, there is no guarantee that information found in any one tertiary resource is accurate or complete.
> - Check the date of publication for print resources or the last update for online resources. There is often a lag time between when information is prepared and when it is published in print. Online tertiary resources might not regularly update all of the information contained, regardless of how frequently they "update."

Similar to obtaining pertinent background information, developing an efficient search strategy is a skill that improves with experience. As stated above, properly classifying the question helps to determine where to begin the search strategy, but each question type or information need might necessitate a completely different search strategy. There are some general guidelines for conducting an appropriate search for information. Most searches begin with the use of tertiary resources. Tertiary resources compile or synthesize information from primary literature such as published clinical studies, and include textbooks, compendia, electronic databases (such as *Micromedex*, *Clinical Pharmacology*), and journal review articles. Tertiary resources provide a broad overview of a topic, which could suffice when the information needed is straightforward and fairly well-documented. Even if a complete answer is not found in the tertiary resource, it will often explain the involved topic or cite other relevant articles that could help tremendously when continuing the search process.

If tertiary resources do not provide a complete answer or if more detailed or current information is needed, secondary resources should be used to locate primary information resources efficiently (see Chapter 3 for a description of secondary resources). It should be noted that secondary resources are also used to find review articles, but again, these are classified as tertiary literature since they only summarize findings from original research and other sources. However, review articles may further expand upon the information found in other tertiary resources and they provide citations for primary literature (e.g., original research and studies) that can be accessed for more detailed information. It takes time and practice to become proficient at searching for information. As a start, clinicians need to become familiar with commonly used and reputable resources. However, it is not enough simply to read about these resources; clinicians must gain practice in using them to learn the most efficient way to find desired information.

Evaluate, analyze, and synthesize information found

Finding information in resources is only part of what is needed to answer a question appropriately. It is imperative that clinicians critically evaluate the information retrieved to ensure that it is valid and meets their needs. In order to critically evaluate information located from the search strategies employed, one must be cognizant of the limitations of each type of resource.

To evaluate properly, several questions should be considered:

1. What is the currency of the resource and the information contained? As previously indicated, tertiary resources often have considerable lag times for the inclusion of new information. Even a newly published edition of a tertiary resource may contain out-of-date information.
2. What is the expertise of the authors? Do the authors have sufficient education and experience to discuss the topics covered by the resource appropriately?
3. Does the resource provide reference citations for the information included? Proper reference citations lend credibility to the information provided and allow the clinician to retrieve the articles cited to obtain more detailed information.
4. Was there any obvious bias or were errors present in the resource? Tertiary references are not free from human error, including incorrect interpretation of information and transcription errors. Authors may also intentionally or unintentionally interject personal bias into their interpretation of studies they review.

The main considerations when evaluating secondary resources primarily involve the appropriate use of the secondary source by the clinician. The following questions should be answered when examining these resources:

1. Was the appropriate secondary resource used?
2. Was any database-specific controlled vocabulary used when conducting the search?

3. Were multiple secondary resources used when conducting the search?
4. How much time does it take for the secondary resource to index new primary literature? If there is a fairly long lag time for new articles to be included, the secondary resource may not contain the most recent information. Refer to Chapter 3 for additional discussion regarding secondary resources.

Evaluating the primary literature is discussed in detail in Chapters 5–12, and those concepts should be applied to this step of the modified systematic approach prior to preparing an appropriate response to an inquiry. For additional information regarding primary, secondary, and tertiary resources, refer to Bogner and Giovenale (2012).

Synthesizing a response must take into account a critical evaluation of the information found. Consideration of the limitations of the resources used is paramount to providing a logical, rational response. This is especially true when conflicting information is located. It is unethical to provide an answer that omits relevant information simply because it might not support the recommendation being made. However, clinicians can and should clearly describe the strengths and limitations of any conflicting information to present the conclusion they believe is best supported by the evidence.

The first step in synthesizing a response is restating (in writing or verbally, as appropriate) the initial question, including any pertinent background information. By restating the question, one can ensure that all of the required information will be included in the response. Objective, balanced, and unbiased information should then be presented in a concise, logical format that allows the recipient to follow and understand the response easily. Finally, any response should end with a brief conclusion and summary of the important aspects of the information presented.

Provide response

Responses to information needs or requests must be provided in a timely manner (within the time frame initially identified), at a level appropriate for the requester (e.g., healthcare provider or patient), and in the format agreed upon (verbal versus written). Good communication skills must also be utilized in this step. The delivery of a verbal response (confidently providing information, speaking clearly, correct pronunciation and use of medical terminology) greatly affects how the information is received. In addition to the use of appropriate communication skills when presenting information, those same skills are important for maintaining dialogue with the requester after the information is received. Such dialogue allows any other questions or concerns to be appropriately addressed at the time the response is given. In this regard, the clinician responder should have anticipated in advance additional questions that may arise and be prepared to address them as well.

Follow-up and document

Follow-up is important to ensure that a response truly provides what is needed, and it allows any additional needs to be identified. Follow-up can be done at the time the response is provided and, if necessary, at a later time as well. For example, an answer might be formulated that is complete and accurate based on background-related data or other information available at the time. However, suppose 2 weeks later an article is published that provides additional insight into the question originally asked. This new information should also be provided to the requester.

Documenting medication and health information questions and responses should be routinely done if such responsibilities are part of a clinician's practice responsibilities, or whenever necessary as part of patient care provision. The level of documentation can vary, but at a minimum should include contact information for the requester, the question asked or information needed (including relevant background information), the response, and the references used in preparing the response.

Key Points

- A modified systematic approach to answering medication and health information questions or finding needed information is a useful tool to ensure that all important aspects are considered.
- Obtaining requester demographics and appropriate background information for the question is essential for determining the actual information need.
- A comprehensive, systematic search strategy should be employed that includes the use of tertiary, secondary, and primary resources as appropriate.
- Properly critiquing information is essential to ensure that it is accurate and relevant to the question or need.
- A balanced and unbiased analysis of the relevant information should be prepared that includes discussion of any limitations and an appropriate conclusion.
- The response should be communicated at an appropriate level.
- Follow-up is important to ensure that the response met the needs and to assess whether any additional information is required.

How to apply to practice

Approach each medication-related question or need using a systematic process that includes:

- thoughtful consideration of related patient needs
- a logical searching process to find relevant information
- critical analysis of the strengths and limitations of the resources used as well as the information obtained
- logical, thorough answers that discuss any conflicting information found, provide a conclusion and summary, and include the references used.

Self-assessment questions

Question 1
What is a potential concern when using a systematic approach to respond to medication- or health-related questions?

Question 2
You are asked the following question by another clinician: "What is the risk of taking lisinopril during pregnancy?" What additional background questions should be asked to clarify this question?

Question 3
A physician has a patient taking insulin. He wants to know what the incidence is of a hypersensitivity reaction to insulin. What additional questions would you ask this physician to clarify the information need further?

Question 4
What are some important considerations with the use of tertiary resources for locating information?

Question 5
Why is it important to present a balanced answer (including any limitations of information found) for a specific question?

Question 6
What are the steps for formulating the response to a medication information question?

Question 7
What is the importance of providing follow-up to the question asked?

Summary

A systematic approach to responding to medication- and health-related questions is important to ensure that responses are consistent, accurate, and thorough, regardless of the involved situation. The steps in this approach should progress fluidly and good communication skills should be maintained throughout the process. This chapter described one version of a modified systematic approach to addressing medication- and health-related questions or related information needs, and it detailed the important concepts to consider with each step.

2

Tertiary resources

Learning objectives

Upon completion of this chapter, you should be able to:
- describe the main features of commonly used tertiary references
- identify advantages and disadvantages of various tertiary references
- identify appropriate references to use for answering a variety of types of medication- and health-related questions.

Introduction

As discussed in Chapter 1, tertiary references are often the first place to start when researching a medication- or health-related question. Tertiary references provide an overview of a topic, without the details found in primary literature (e.g., clinical studies). Even if the question is not completely answered by these references, tertiary resources will usually provide the basis for a more thorough information search.

There are many commonly used tertiary references that provide valuable information for healthcare professionals. General information sources are typically those that provide a broad base of material. Some of these references (often referred to as compendia) consist of a series of monographs (e.g., drug descriptions), and these monographs typically follow a standard format. In the case of drug information resources, while the format might markedly differ among resources, the standard format within a given reference allows clinicians to find similar information easily for any medication included. In contrast, other tertiary resources are designed to answer specific types of questions. These resources have a much narrower focus and will provide more detailed information about a specific topic. For example, even though many general drug information resources provide some information about a medication's use in

> **Key Point**
>
> Assess the question being asked to determine the types of resources to use and the level of information required to locate an adequate answer. If the question is general in nature, general information tertiary sources may provide adequate information. If the question focuses on a specific topic, utilize specific resources that address the topic. Often, a combination of general and specific resources will provide the most complete answer to a question.

pregnancy, this information might be broad in nature and not provide much detail beyond the assigned pregnancy category. However, an information-specific reference such as *Drugs in Pregnancy and Lactation* (Briggs et al. 2011) provides a detailed description of a medication's use during pregnancy, and this would allow for a more complete response to this type of question or information need.

The resources discussed in this chapter use various methods for content delivery. Some are available in print only, others are available in print and electronically, and still others are available electronically only. Many times the electronic versions of the references will provide additional information or tools for the clinician. Descriptions follow for several commonly used tertiary resources that are valuable for locating drug- and health-related information. For additional information regarding tertiary resources, refer to Bogner and Giovenale (2012).

General drug information resources

AHFS Drug Information (American Society of Health-System Pharmacists)

- Available in print and electronic formats
- Consists of an extensive collection of drug monographs that undergo a comprehensive review process prior to publication
- The monographs are arranged by the AHFS Pharmacologic-Therapeutic Classification, which consists of a numeric classification system to allow for easy review of groups of drugs with similar activities
- While each drug has a unique monograph, there are also general statements on groups of drugs (e.g., salicylates) that belong to the same pharmacologic or chemical class. The general statements precede the individual drug monographs, and both the general statements and individual drug monographs should be consulted when gathering information
- Monograph sections include: title and synonyms; introduction; uses (labeled and off-label); dosage and administration; cautions; drug interactions; laboratory test interferences; acute toxicity; chronic toxicity; pharmacology; mechanism of action/spectrum/resistance (for anti-infectives); pharmacokinetics; chemistry and stability; and preparations
- Information included in AHFS is thoroughly referenced. However, access to referenced statements within a monograph can only be gained through electronic versions.

Clinical Pharmacology (Elsevier/Gold Standard)

- Available electronically
- Covers a wide range of prescription, over-the-counter (OTC), investigational, nutritional, and herbal products

- Information is peer-reviewed and updated using a real-time format (allows for the inclusion of late-breaking medical news, trials, and other information). Information is based on products and practices within the USA
- Monograph sections include: description/classification; mechanism of action; pharmacokinetics; indications/dosage (labeled and off-label); administration; contraindications/precautions; adverse reactions; interactions; how supplied (includes product-specific information including manufacturer, formulations, and inactive ingredients); monitoring parameters; and IV compatibility
- Abbreviated monographs are used for older drugs, drugs not widely available, and investigational drugs
- Includes a drug identification tool, as well as the ability to run reports for drug interactions, IV compatibility, product comparisons, and clinical comparisons
- References are cited for controversial or less well-known concepts only.

Drug Facts and Comparisons (Wolters Kluwer Health)
- Available in loose-leaf (updated monthly) and bound (updated annually) print versions and electronic formats
- Covers prescription, OTC, investigational, and orphan drugs
- Drugs are divided into related therapeutic or pharmacological groups so that similar drugs can be easily compared
- Monograph sections include: pharmacology; indications; administration and dosage; pharmacokinetics; clinical evidence; interactions; adverse reactions; warnings; pregnancy/lactation; overdose information; and references
- Includes more than 3000 comprehensive charts and tables providing comparisons of products with similar formulations, manufacturer's contact information, laboratory values, and pharmaceutical abbreviations
- Electronic version includes advanced dosing tools and drip rates for life support medications, clinical calculators, drug identification and interaction tools, and the ability to search for drugs by disease/symptoms.

Drug Information Handbook (Lexicomp)
- Available in print and electronic formats
- Useful as a quick reference, but often the information is not as detailed as other compendia
- Monographs are listed alphabetically under generic drug names. Brand names are included in the alphabetical listing and are cross-referenced to the generic monograph
- Monograph sections include: pronunciation of generic drug name; US and Canadian brand names; pharmacologic category; use (labeled and

off-label); pregnancy/lactation; contraindications; warnings; adverse reactions; interactions; mechanism of action; pharmacodynamics/ kinetics; dosage/administration; patient information; dosage forms; and, if available, extemporaneous compounding formulas
- Includes several appendices with useful clinical information such as abbreviations and measurements, comparative drug charts, cytochrome P450 inducers and inhibitors, and toxicology information
- Internet version includes tools for drug identification, interactions, IV compatibility, and various clinical calculators. Online versions can also be purchased to include all of the *Lexi Drugs Handbook* products such as the *Pediatric and Neonatal Dosage Handbook* and the *Geriatric Dosage Handbook*, in addition to other Lexicomp references including *Lexi-Tox* (toxicology information), *Lab Tests and Diagnostic Procedures*, and *Lexi-Patient Education* resources.
- Internet version can also incorporate *AHFS Drug Information* along with Lexicomp resources (the inclusion of *AHFS Drug Information* adds a more comprehensive and detailed source of medication information to the online database).

Martindale: The Complete Drug Reference (Pharmaceutical Press)
- Available in print and electronic formats
- International scope of information (contains proprietary preparations from over 41 countries and regions) and excellent resource for locating products available in other countries
- Includes information on select herbals, diagnostic agents, radiopharmaceuticals, pharmaceutical excipients, toxins, and poisons in addition to drugs and medicines
- Separated into sections containing substance monographs; preparations; directory of manufacturers; and multilingual pharmaceutical terms
- Substance monographs are placed into chapters based on groups of drugs with similar actions or uses. The introductions of each chapter may also include disease treatment reviews for drugs contained within the chapter
- The last chapter in the monographs section contains a series of monographs, in alphabetical order, of drugs or substances not easily classified or otherwise included in the previous chapters
- Monograph sections include: atomic and molecular weights; pharmacopoeias containing the substance; pharmaceutical information (solubility, storage, etc.); pharmacological and therapeutic information (adverse effects, precautions, interactions, pharmacokinetics, and uses and administration)
- Preparations section contains the proprietary product name, the manufacturer or distributor, the active ingredients, and a summary of indications as provided by the manufacturer

- Information provided is based on published data and is extensively referenced.

Micromedex (formerly Thomson Reuters Healthcare, now Truven Health Analytics)

- Available electronically
- Compilation of several databases with specific types of information (the specific databases available depend on the subscription package obtained)
- Databases include: Drugdex, Diseasedex, Alternative Medicine, Detailed Drug Information for the Consumer, Identidex, Index Nominum, Lab Advisor, MSDS from USP, P&T QUIK Reports, Physicians Desk Reference, Poisondex, Reprorisk, TOMES, Martindale, Red Book Online, United Kingdom Drug Information, Italian Drug Database, Formulary Advisor, Neofax, Pediatrics, and CareNotes
- Includes several tools for clinicians such as drug interactions; IV compatibility; drug identification; toxicology and drug product lookup; drug comparison; and several clinical calculators that include antidote dosing and nomograms; laboratory values; dosing tools; and clinical/measurement calculators
- Drugdex is the proprietary product database consisting of drug monographs. It can be viewed as a summary document or as a detailed document. The detailed document should be utilized for locating comprehensive information
- Monograph sections include: overview; dosing information (drug properties, storage and stability, adult dosing, pediatric dosing, specialty population dosing); pharmacokinetics (onset/duration, drug concentration levels, ADME); cautions (black box warnings, contraindications, precautions, adverse reactions, pregnancy/breastfeeding, drug interactions); clinical applications (monitoring parameters, patient instructions, place in therapy, pharmacology, therapeutic uses, comparative efficacy/evaluation with other therapies); and references
- Incorporates Food and Drug Administration (FDA) -approved (labeled) uses and unlabeled uses. The description of unlabeled uses for drugs is comprehensive and thoroughly referenced to available literature
- Provides an international scope of information if *Index Nominum* or *Martindale* is included in the subscription package
- Reprorisk is a collection of four databases (Reprotext, Reprotox, Shepard's, and Teris) that together provide a comprehensive discussion of reproductive risks (teratogenicity, reproductive risks from acute and chronic exposure, and reproductive risk from male drug exposure) associated with drugs or other substances.

Physician's Desk Reference (PDR) (PDR Network)
- Available in print and electronic formats
- Compilation of FDA-approved product labeling for commonly prescribed proprietary drugs
- Monograph (product label) sections include: indications and usage; dosage and administration; contraindications; warnings/precautions; adverse reactions; drug interactions; use in specific populations; overdosage; description; Clinical Pharmacology; nonclinical toxicology; clinical studies; how supplied/storage and handling; patient counseling information
- Does not include any investigational or off-label uses of medications
- Important to note that the PDR is paid by various manufacturers to present their product's information in the PDR
- Also contains a manufacturers' index, a generic/brand cross-reference table, dietary supplements section, and color images of select medications
- *PDR for Herbal Medicines*, *PDR for Nonprescription Drugs*, *PDR for Ophthalmic Medicines*, and *PDR for Nutritional Supplements* are also available.

Information-specific resources
Adverse reactions

Litt's Drug Eruptions and Reactions Manual *(Informa Healthcare)*

> **Key Point**
>
> Many general drug information resources list labeled and unlabeled uses for drugs, but *AHFS Drug Information* and the *Micromedex Drugdex* database provide *detailed* descriptions for unlabeled uses.

- Provides information on adverse reactions, drug eruptions, and clinically relevant drug–drug interactions for over 1000 drugs
- Includes: A–Z drug profiles; description of important reactions; drugs that cause important reactions; and an index of trade names, synonyms, and generic names
- Includes prescription, OTC, biologic, and supplement products.

Meyler's Side Effects of Drugs *(Elsevier)*
- Comprehensive source of adverse drug reactions
- Uses an encyclopedic format with drug monographs organized alphabetically
- Each monograph contains detailed information about a drug's adverse effects with comprehensive references to the primary literature and reports of clinically relevant drug interactions

- *Side Effects of Drugs Annual* editions are published yearly and supplement the information about adverse drug reactions between new editions of the main Meyler's editions (both the encyclopedic volume of Meyler's and subsequent annuals should be utilized for a complete search).

Compatibility and stability of parenteral products

Handbook on Injectable Drugs *(ASHP)*
- Collection of monographs summarizing pharmaceutics and compatibility information for approximately 350 parenteral medications
- Monographs are presented alphabetically by nonproprietary product name
- Monograph sections include: products; administration (includes route and rate of administration); stability; compatibility information (in tabular form and includes information for solutions, admixtures, drugs in syringe, and y-site administration); additional compatibility information; and other information
- Also contains information regarding parenteral nutrition formulas in an appendix.

King Guide to Parenteral Admixtures *(King Guide Publications)*
- Collection of monographs summarizing the compatibility of nearly 500 parenteral products
- Includes limited information regarding product descriptions and stability
- Provides tables indicating compatibility in 12 common fluids when a drug is in solution or mixed with other drugs
- Includes information for parenteral nutrition that allows for comparisons of compatibility of similar parenteral nutrition admixtures.

Drug interactions

Drug Interaction Facts *(Wolters Kluwer Health)*
- Includes drug–drug and drug–food interactions
- Fully referenced and interactions are indexed by generic name, brand name, medication class
- Each interaction monograph includes: significance rating; onset; severity rating; documented effects; mechanisms; and management options.

Drug Interactions Analysis and Management *(Wolters Kluwer Health)*
- Drug interaction monographs arranged alphabetically
- Each monograph includes: significance classification; summary; risk factors; mechanism; clinical evaluation; related drugs; management options; references
- Management options are emphasized to improve patient outcomes.

Evaluations of Drug Interactions *(First Databank)*
- Includes prescription and OTC drug interactions
- Monographs of interactions are divided by drug class
- Monograph sections include: title (generic name); severity level; summary; related drugs; mechanism; recommendations; references; interaction summary box
- Endorsed by the American Pharmaceutical Association (APhA) and reviewed by an APhA scientific panel.

Stockley's Drug Interactions *(Pharmaceutical Press)*
- Includes interactions between drugs, herbal medicines, foods, drinks, pesticides, and drugs of abuse
- Contains concise and fully referenced monographs of interactions
- Monograph sections include: introductory summary of the interaction; clinical evidence for the interaction; an assessment of the clinical significance of the interaction; and practice management guidance for the interaction
- Both British and American drug names are included.

Extemporaneous compounding

A Practical Guide to Contemporary Pharmacy Practice *(Lippincott Williams & Wilkins)*
- Concise handbook that covers all steps of compounding from receiving the prescription to preparing and completing the compound
- Contents include: processing the prescription (includes beyond use dating); calculations; drug preparations; excipients; nonsterile dosage forms; sterile dosage forms; veterinary pharmacy; compatibility and stability.

Art, Science, and Technology of Pharmaceutical Compounding *(APhA)*
- Comprehensive review of compounding beginning with necessary facilities and equipment, and progressing through the process of compounding various products and dosage forms, including sterile products
- Covers additional topics related to compounding such as veterinary compounding, compounding for special populations and procedures, compounding for clinical studies, compounding with hazardous drugs, and compounding in the event of a natural disaster or terrorist attack.

Extemporaneous Formulations for Pediatric, Geriatric, and Special Needs Patients *(ASHP)*
- Provides a brief introduction to the practice of compounding medications
- Gives compounding formulas for 160 medications that can be used for patients unable to take traditional dosage forms

- Each formula includes: ingredients; preparation details; storage conditions; special instructions; expiration date; and references.

Pediatric Drug Formulations *(Harvey Whitney Books)*
- Provides a brief introduction to compounding
- Includes compounding formulas for commonly used medications
- Each formula includes: dosage form; made from; concentration; stability; reference; storage; label information; ingredients; notes; and instructions.

Trissel's Stability of Compounded Formulations *(APhA)*
- Compilation of over 400 product monographs arranged alphabetically by generic drug name
- Monographs are divided into three sections: properties (characteristics of the drug including solubilities, pH, and osmolality); general stability considerations (recommendations for packaging and storage); and stability reports of compounded preparations (summarizes published literature regarding stability of different formulations)
- Includes appendices that discuss beyond-use dating of both sterile and nonsterile products
- Fully referenced in each section.

Herbal, natural, or alternative medicines

Natural Medicines Comprehensive Database *(Research Faculty)*
- Comprehensive collection of monographs covering herbal and other complementary medicines
- Each monograph is evidence-based and fully referenced
- Monograph sections include: synonyms; scientific name; uses; safety; effectiveness; mechanism of action; adverse reactions; interactions (includes other herbals, drugs, food, lab tests, and diseases); dosage/administration; and editor's comments
- Provides guidance for use based on available evidence for safety and efficacy
- Includes several charts such as therapeutic efficacy; interactions between drugs and natural medicines; and most popular commercially available products
- Online version allows users to search by substance brand name, synonyms for the substance, disease state, or conditions.

Natural Standard *(Natural Standard; Mosby, an affiliate of Elsevier – print versions)*
- In print and electronic formats, but electronic version includes more information with better search functionality

- Provides evidence-based information for herbal products as well as other complementary and alternative medicines
- Each product is graded (A–F) to reflect the quality of scientific evidence for or against the use of the product for a specific therapeutic use
- Monograph sections include: synonyms/common names; clinical bottom line/effectiveness (includes safety discussion); dosing/toxicology; precautions; interactions; mechanism of action; history; evidence table (summarizes published literature); evidence discussion; products studied; author information; and references
- Electronic version includes several charts and tables and other useful tools such as calculators, patient handouts, and nutrition labels.

The Review of Natural Products *(Wolters Kluwer Health)*
- Compilation of over 350 evidence-based monographs of natural products (not as comprehensive as *Natural Medicines Comprehensive Database* or *Natural Standard*)
- Monograph sections include: botany; history; chemistry/pharmacology (discussed in more detail than other herbal references); medicinal uses; toxicology; patient information (brief summary); and significantly documented drug interactions
- Includes a list of natural product websites, a list of herbal diuretics, a mushroom poisoning decision chart, and a discussion of the use of natural products during pregnancy and lactation.

International drug products

Index Nominum: International Drug Directory *(MedPharm)*
- Available in print and electronic formats
- Edited by the Swiss Pharmaceutical Society
- International pharmaceutical reference of medications, proprietary names, synonyms, chemical structures, and therapeutic classes of substances from manufacturers representing 171 countries
- Limited therapeutic information is provided for each entry (usually only the therapeutic class is identified with no other content)
- Provides manufacturer's contact information for each product
- Useful to identify international drug products by trade names, generic names, or synonyms.

Martindale: The Complete Drug Reference *(Pharmaceutical Press)*
- See previous discussion of *Martindale* in General drug information resources section
- Useful resource for identifying foreign drugs: it includes more information about each drug than *Index Nominum*.

Pediatric dosing

BNF for Children *(BMJ Publishing Group, RCPCH Publications, and Pharmaceutical Press)*

- Based on the *British National Formulary* and compiles concise monographs that provide dosing information for ages ranging from neonates to adolescents
- Included material is from emerging evidence, best practice guidelines, and clinical experts
- Monograph sections include: cautions; contraindications; hepatic impairment; renal impairment; pregnancy; breastfeeding; side-effects; licensed use (UK); indication and dose; administration; approved name; and proprietary name
- Monographs are organized by therapeutic topics
- An introduction to the general treatment of each therapeutic topic precedes each section, followed by the individual monographs for drugs used to treat that condition.

Harriet Lane Handbook *(Mosby, an affiliate of Elsevier)*

- Concise handbook with discussions of diagnosis and treatment of diseases and conditions seen in pediatric patients
- Contains a drug formulary section that provides concise dosing information for common medications used in pediatric patients
- Monograph sections in the drug formulary include: product information (generic/trade names, how supplied, therapeutic category); drug dosing; brief remarks (side-effects, drug interaction, precautions, monitoring, other relevant information)
- Drug monographs are brief and do not provide detailed information or references.

Pediatric & Neonatal Dosage Handbook *(Lexi-Comp)*

- Compilation of over 900 individual monographs focused on pediatric and neonatal dosing
- Monographs are organized alphabetically by generic drug name
- Monograph sections include: medication safety issues; related information; US and Canadian brand names; therapeutic category; uses; pregnancy/lactation; contraindications; warnings/precautions; adverse reactions; interactions; mechanism of action; pharmacodynamics/kinetics; dosing/administration; patient information; dosage forms; limited references
- Includes over 100 extemporaneous preparation recipes in applicable monographs
- Includes a comprehensive appendix with multiple types of clinical information including acute life support information, comparative drug

charts, conversion tables, growth and development information, and toxicology.

Pharmacy practice

Remington: The Science and Practice of Pharmacy *(Pharmaceutical Press)*
- Definitive text describing the science and practices associated with the profession of pharmacy
- Covers a broad range of topics related to pharmacy, including: orientation (to pharmacy practice); pharmaceutics; pharmaceutical chemistry; pharmaceutical testing; manufacturing; pharmacokinetics/dynamics; pharmaceutical and medicinal agents; and pharmacy practice
- Future editions will be available in an online format.

Pregnancy and lactation

Drugs in Pregnancy and Lactation *(Lippincott Williams & Wilkins)*
- Compilation of 1200 common drugs organized alphabetically by generic drug name
- Summarizes currently available literature describing fetal risk with drugs and the impact of those drugs during breastfeeding
- Monograph sections include: generic US name; pharmacologic class; risk factor; pregnancy summary; fetal risk summary; breastfeeding summary; and references
- Pregnancy summary provides a brief review of the information presented.

Lact-Med *(National Library of Medicine)* – refer to Chapter 4 discussion of *Toxnet databases*

Medications and Mother's Milk *(Hale Publishing)*
- Compilation of more than 900 drugs and vaccines arranged alphabetically
- Monograph sections include: trade names (US and international); uses; American Academy of Pediatrics (AAP) recommendation; synopsis of use in breastfeeding; pregnancy risk category; lactation risk category; adult concerns; pediatric concerns; drug interactions; theoretic infant dose; relative infant dose; adult dose; alternatives; select pharmacokinetic data; and references (limited)
- Chemotherapeutic agents and radiopharmaceutical use during lactation, drugs and herbals contraindicated during lactation, commonly used cold remedies, and other clinical information are discussed in appendices.

Reprorisk *(Micromedex – refer to previous discussion under General drug information resources)*

Therapeutics

Applied Therapeutics: The Clinical Use of Drugs *(Lippincott Williams & Wilkins)*

- Uses a case-based approach to illustrate the fundamentals of therapeutics
- Divided into chapters organized by therapeutic topics
- Each chapter provides an overview of therapeutic information (pathophysiology, diagnosis, treatment), followed by specific cases with directed questions to illustrate specific clinical scenarios. Each direct question is followed by a detailed comprehensive discussion of the answer.

Goodman and Gilman's The Pharmacological Basis of Therapeutics *(McGraw-Hill)*

- Focused on the pharmacology of drugs; links discussions of the pharmacology to the pharmacodynamics and therapeutic use of drugs
- Divided into chapters organized by the body system affected by the drugs discussed
- Contains information on physical and chemical properties, pharmacology, mechanism of action (details structure–activity relationships), adverse effects, toxicology, and therapeutic use of drugs.

Pharmacotherapy: A Pathophysiological Approach *(McGraw-Hill)*

- Provides detailed discussions of pathophysiology and treatment of various disease states and conditions
- Organized into chapters divided by body systems
- Chapters contain information regarding epidemiology, pathophysiology, clinical presentation of disease, treatment, evaluation of treatment and economic outcomes, with references
- Key concepts are listed at the beginning of each chapter and their corresponding numbers are highlighted in the text of the chapter
- Clinical controversy boxes are included to highlight important points when there are conflicting data regarding the therapeutic information.

Pharmacotherapy Principles and Practice *(McGraw-Hill)*

- Details management of various disease states and conditions using condensed, important information from *Pharmacotherapy: A Pathophysiological Approach*
- Chapters contain material discussing epidemiology, etiology, pathophysiology, treatment, and outcome evaluation, with references

- Key concepts are listed at the beginning of each chapter and their corresponding numbers are highlighted in the text of the chapter
- Patient encounters (cases) are used in the chapters to illustrate additional points or to reinforce material presented in the text.

Key Points

- It is appropriate to consult tertiary resources first when answering a drug information question or finding needed medicine- or health-related information to provide an overview of a topic or to provide references that can be accessed for more detailed information.
- Consult appropriate tertiary resources based upon the question being asked or type of information needed – do not rely upon general tertiary information resources to find specific, focused types of information.
- If possible, do not rely on information found in only one tertiary resource. Always strive to verify the information found in one source with other available resources to ensure information completeness and accuracy.

Self-assessment questions

Question 1
A physician would like information about the use of paroxetine to treat a postmenopausal patient suffering from hot flashes. The patient complains of multiple nighttime awakenings because of the hot flashes as well as multiple episodes during the day. Which tertiary references would be best suited to begin researching this question?

Question 2
A patient tells the pharmacist that she is trying to become pregnant. She currently takes topiramate for migraine prophylaxis. She is aware that topiramate has a "poor" pregnancy rating but asks about the specific risks and effects of taking topiramate if she becomes pregnant. Which tertiary reference(s) would be best suited to begin researching this question?

Question 3
The same patient from Question 2 returns to the pharmacy a year later. She stopped taking the topiramate for migraine prophylaxis prior to becoming pregnant. She delivered a healthy full-term infant 3 months ago and she is currently breastfeeding. Since the delivery, her migraine headaches have increasingly worsened in frequency and severity. She would like to begin topiramate for migraine prophylaxis since this previously controlled the frequency and severity of her migraine attacks. Before she talks to her physician, she would like to know what the risk would be if she continued to breastfeed her infant while taking topiramate. What sources would be best suited to begin researching this question?

Question 4
A patient asks his physician about the use of cinnamon to help control his blood sugar: he read about this in a lay publication. The physician asks the pharmacist for help in evaluating any available information about the use of cinnamon to control blood glucose in diabetes. Which resource would be best suited to begin researching this question?

Question 5
A patient presents a prescription to the pharmacy for baclofen 5 mg/mL oral suspension. The pharmacist needs to find a compounding formula to prepare the product and fill the prescription. Which resources would be best suited to begin locating this information?

Question 6
A patient states he has recently returned from an extended stay in Japan as part of a work assignment. While in Japan, he was evaluated for hypertension and placed on a medication called Almylar. The patient would like to know if that medication is available in the USA and, if not, what would be an alternative medication to take? Which resources would be best suited to begin researching this question?

Summary

It is important for healthcare professionals to know the types of information included in the various tertiary resources, as well as their strengths and limitations, so they can have on hand those references that would be most useful for their practices. Such knowledge will also help to improve efficiency when using tertiary resources since it allows clinicians to go directly to the best source(s) of information to find what is needed. This chapter focused on the important characteristics of many commonly used tertiary resources for locating medication- and health-related information.

3

Secondary resources

Learning objectives

Upon completion of this chapter, you should be able to:
- describe the main features of commonly used secondary resources
- identify proper search strategies for accessing secondary resources
- identify which secondary resources to utilize for specific types of information.

Introduction

Secondary resources are used by clinicians to locate clinical studies or other original research (referred to as the *primary literature*) that are published in medical or health-related journals. The amount of primary literature published each year continues to grow and has reached a staggering number. Thus, it is impossible to keep abreast of all the new primary literature published even if pared down to specific topics needed for a practice setting. The good news is that clinicians do not have to track every new article published; one simply needs to know how to use secondary resources to find pertinent articles when needed.

> **Key Point**
> - The overall number and types of journals indexed by different secondary resources can vary greatly. For example, Medline indexes approximately 18 million references from over 5500 sources. To ensure a complete search for primary literature, multiple secondary resources should be used.

Secondary resources are generally categorized as *abstracting* or *indexing* services. Indexing services provide a bibliographic citation (authors, title, journal name, volume, pages, year) for each journal article included. Citations allow clinicians to retrieve articles but do not provide any additional content information. Abstracting services, in addition to citations, provide a brief description (abstract) of the content of each article. An abstract allows clinicians to determine if the specific article meets their needs before accessing the complete resource.

One obstacle frequently encountered when using secondary resources is the availability of the complete (full-text) journal article for the citations/abstracts retrieved. Some full-text articles will be available free online or are in commonly read journals and thus can be obtained fairly easily.

Other citations may be from more obscure journals/sources or from international journals. In these cases, the article may be hard to obtain, especially if a clinician is not affiliated with or does not have ready access to a medical library. The full text of some citations may be obtained directly from the journal's publisher for a fee. If the citation is from Medline, the Loansome Doc program can be used to order the full text of a citation for a small charge. This chapter provides an overview of several commonly used secondary resources.

Searching secondary resources

Choosing the correct searching terms is key to performing efficient and effective searches of secondary resources. The majority of these resources will use *Boolean operators* (AND, OR, and NOT) to combine search terms and either broaden or limit the results obtained. When using AND to combine search terms, the search is narrowed because only results that contain *all* of the terms are obtained. The Boolean operator OR will broaden the search because results that contain *any* of the search terms are obtained. The Boolean operator NOT will also limit the results obtained, but should be used cautiously because the results may be narrowed too much and relevant articles may be missed.

Boolean operators allow the clinician to control the specificity of the search results. A properly executed search strategy using appropriate Boolean operators can often pinpoint the most relevant articles or at least minimize the number of irrelevant citations returned. However, an improperly conducted search could result in either many missed articles or too many irrelevant citations. Suppose a clinician wishes to find citations discussing the use of enalapril or propranolol to treat hypertension. One might be tempted to enter the following into the search box of an electronic secondary resource: "hypertension AND propranolol OR enalapril." However, keep in mind that the search terms will usually be searched from left to right in the order found. The secondary resource will return citations for articles that contain BOTH "hypertension" and "propranolol," plus ALL articles that contain the term "enalapril," regardless of whether "hypertension" or "propranolol" was present. This would return a large number of irrelevant citations that would obscure the ability to identify the truly relevant articles. To avoid this problem, parentheses should be used around any terms that are combined using the OR operator to group those terms together effectively during the search. The proper way to conduct the search would be: "hypertension" AND ("propranolol OR enalapril") – this would result in the retrieval of all articles containing "hypertension" that also contain either "propranolol" and "enalapril."

Worked example

Example 3.1

A clinician is using a secondary information database to find information regarding the combined use of aspirin and ticlopidine or the combined use of aspirin and clopidogrel to prevent clotting after placement of a coronary stent. *What is the best way to combine these drug terms in the search box?*

Answer

To combine the terms above properly, parentheses should be used to group the items combined with the OR operator. A good search strategy would be to group the terms as follows: "aspirin" AND ("ticlopidine" OR "clopidogrel"). This combination of terms would result in citations that include aspirin plus *either* ticlopidine or clopidogrel.

Some secondary resources (Google Scholar) allow users to search using phrases or multiple terms without Boolean operators and are similar to general internet search engines. While this may be more intuitive for the average clinician performing the search, the specificity and precision that can be gained from properly using Boolean operators in the search function are lost.

Key Point

The following are important in order to search secondary resources effectively:

- Properly use Boolean operators to combine search terms.
- Form the research question and select the key concepts that should be included in the search strategy.
- Become knowledgeable about the controlled vocabulary or thesaurus that each resource uses for indexing search terms and incorporate those terms when appropriate.
- Consider whether plurals or other possible endings for a word should be part of the search and use the proper truncation symbol for such words.
- Use commonly used synonyms for search terms whenever appropriate.

It is also important to identify the exact search terms to use when accessing secondary resources. First, examine the question to determine which parts should be included as search terms. To do so, ask the following: Which key terms from the question are necessary to ensure that relevant articles are retrieved? In the worked example above, important terms include ticlopidine, clopidogrel, aspirin, clotting, and coronary stent. If "coronary stent" is excluded, one might obtain citations for the use of these drugs to prevent clots from other causes that might not extrapolate to a coronary stent.

A further searching challenge when using secondary resources is that most do not use the same indexing terms for the articles included. For instance, the secondary resource Medline uses Medical Subject Headings (MeSH) to index articles, the Iowa Drug Information Service (IDIS) uses the United States Adopted Name to index drug terms and the *International Classification of Diseases* to index nondrug terms, while International Pharmaceutical Abstracts

(IPA) uses a completely different terminology system (descriptors) to index citations. Although there tends to be similarity in the indexing of drug names between resources, there can be much variability in the indexing of nondrug terms. The terminology a secondary resource uses when indexing terms is referred to as the *controlled vocabulary* for that specific resource. It is important to understand how each secondary resource indexes topics and to use the controlled vocabulary for each key term in the involved question to locate relevant citations efficiently and effectively. However, even when the index terminology is known and used, some topics will not have a unique index term. In those instances, clinicians must use "free text" or their own word choices in the search strategy. Often, it is a combination of index terms and free-text words that provides the most effective search when specific index terms are not available.

Another potential area of confusion when searching secondary resources involves the use of plural terms and synonyms. Using the plural form of a search term could markedly change the search results depending on how the terms are indexed in the secondary resource. It may be prudent to use both the singular and plural forms of a search term or use truncation (often done by placing an asterisk at the end of the word, although this can vary depending on the secondary database) to search effectively. Truncation is useful because it will include other possible endings for a word in addition to plurals. For example, placing a truncation symbol after pharm* will result in the inclusion of "pharm*acy*," "pharm*acies*," "pharm*acist*," "pharm*acists*," or "pharm*acology*." Similarly, clinicians should think of possible synonyms for certain terms that could be indexed multiple ways. For example, the terms "adverse effects," "side-effects," and "toxicity" could all be used to describe a similar topic. Depending on the specific secondary resource, any or all of those terms could be used to obtain relevant articles, thus, it may be prudent to use multiple synonyms in a search strategy.

Finally, most secondary resources allow clinicians to narrow their search results by using *limits*. Limits allow the searcher to focus the results further to specific types of citations. For instance, if looking for clinical trials evaluating the use of a drug, a clinician could place a limit on the search so that only clinical trials are obtained in the search results. This eliminates any unwanted review articles, case reports, editorials, or other irrelevant citations from populating the search retrieval. Common examples of searching limits that can be placed on searches include: age ranges, full-text articles, type of publication (review articles, clinical trials, meta-analysis), language, human versus animal, and date of publication. Limits should be used cautiously, however, since one might miss retrieving citations that would otherwise have been relevant. It may be judicious to run an initial search without using limits and, if too many articles are retrieved, then to place necessary limits on the search.

Commonly used secondary resources

A brief description follows of several secondary resources useful to healthcare professionals.

CINAHL

- Produced and distributed by EBSCO Publishing.
- Cumulative index to nursing and allied health literature.
- Available in different versions. CINAHL is the basic database that indexes over 3000 journals and provides full-text articles for 71 journals. CINAHL with Full Text also indexes over 3000 journals and provides full-text articles for over 600 journals. CINAHL Plus indexes nearly 5000 journals and provides full-text articles for 79 journals. CINAHL Plus with Full Text also indexes nearly 5000 journals and provides full-text articles for nearly 800 journals.
- CINAHL covers all aspects of nursing, and also includes topics such as biomedicine, health information management, alternative and complementary medicine, consumer health, occupational therapy, health promotion and education, and many others related to allied health.
- CINAHL subject headings are the controlled vocabulary used to index literature in this database. These terms were developed to reflect the terminology used by nursing and allied health professionals. CINAHL subject headings follow a similar hierarchical structure as MeSH terms used in Medline (see Medline, below).
- CINAHL also offers access to select healthcare books, nursing dissertations, standards of practice, and selected conference proceedings.

Cochrane Library

- Produced by the Cochrane Collaboration and published by John Wiley.
- Collection of six databases that include the Cochrane Database of Systematic Reviews, the Database of Reviews of Effectiveness (DARE), and the Cochrane Central Register of Controlled Trials.
- Cochrane Database of Systematic Reviews includes all of the Cochrane reviews. Cochrane reviews are peer-reviewed, systematic reviews of specific clinical topics. These reviews critically analyze the available studies on the topic and review and summarize the findings. If the clinical trials are similar, then a meta-analysis of the data may also be conducted.
- DARE contains abstracts of systematic reviews that have undergone quality assessment. The abstracts include a summary of the review and a critical appraisal and commentary of the overall quality of the review.

- The Cochrane Central Register of Controlled Trials contains details of published trials from bibliographic databases (mainly PubMed and Embase).
- The reviews provided in the Cochrane Library are extensively reviewed and must follow a systematic approach so there is consistency among the reviews. This structured approach also minimizes bias and provides reliable findings to be used to guide clinical decisions.

Embase

- Produced by Elsevier.
- Embase has a broad biomedical scope with coverage of basic biomedical science, veterinary science, allied health topics, pharmacology, pharmaceutical science, and clinical research.
- Embase includes all of the journals and citations covered by Medline, plus over 2000 additional journals not indexed in Medline.
- Embase uses a controlled vocabulary from Elsevier's life science thesaurus known as Emtree. Emtree is a hierarchical format of index terms that includes chemical names, trade names, and research codes that are mapped to generic drug names, and it also has a listing of disease terms. These terms can be found within the Embase database by using the drug or disease search functions in Embase.
- Emtree also contains all of the MeSH terms used in Medline.

Google Scholar

- Produced by Google.
- Allows clinicians to search scholarly literature at no charge.
- Uses an internet search engine style that allows users to enter simple terms or phrases and the search engine determines the significance of the results. However, this approach also loses some of the specificity and relevance of results that can be obtained when using appropriate Boolean operators and a controlled vocabulary.
- Limitations of Google Scholar include a lack of information about the journals covered, so it is difficult to ascertain how complete the search results are. Additionally, Google Scholar ranks the relevance of articles by various methods, including the citation counts for the article. Thus, new articles may not appear as relevant in the search results because they do not have the citation counts that older articles might have.

International Pharmaceutical Abstracts (IPA)

- Produced by Thomson Reuters and available through numerous vendors.
- IPA focuses on pharmacy and pharmacy practice.

- Indexes over 750 journals and almost 600 of those are pharmacy-related. IPA also indexes many state and local pharmacy publications, some foreign pharmacy journals, and abstracts from meetings of professional pharmacy organizations such as the American Society of Health-System Pharmacists (ASHP). Including these sources broadens the scope of IPA's coverage, but also can make obtaining the full text of the citations difficult.
- Available in print and electronic formats. The electronic format contains additional citations not included in the print version. The print version is more difficult and time-consuming to search.
- Topics covered include: biopharmaceutics and pharmacokinetics, new drug delivery systems, pharmacist liability, legal, political, and ethical issues, pharmacology, drug stability, and drug evaluations.
- Abstracts for clinical studies include information regarding the study design, number of patients studied, and the dosages, dosage forms, and dosage schedules of the drugs involved.
- IPA uses a variety of index terms (descriptors) for referencing topics. To conduct a complete search in IPA, multiple synonyms for each term being searched must be used.
- All abstracts are presented in English, even if the source of the citation is a foreign journal.

Iowa Drug Information System (IDIS)

- Produced by the University of Iowa, Division of Drug Information Service.
- Indexes articles from over 200 medical and pharmaceutical journals covering a broad range of topics. IDIS provides the full-text article for every indexed citation.
- Articles are only indexed if they are directly relevant to drug use in humans. This narrows the focus of the indexed results obtained from IDIS.
- IDIS uses a controlled vocabulary to index topics. Drug terms are indexed using the United States Adopted Name and a seven- or eight-digit modified AHFS number. Disease terms are indexed using the *International Classification of Disease* names and code numbers. Additional terms (descriptors) can also be indexed. Descriptors include article type, outcomes, pharmaceutics, administration, drug use, and side-effects.
- To help manage the controlled vocabulary, IDIS includes an electronic thesaurus tool that will provide the controlled vocabulary term for any topic entered.
- IDIS also indexes Food and Drug Administration (FDA) approval packages. FDA approval packages are a compilation of all information

received by the FDA during the drug approval process. IDIS prepares a table of contents for the FDA approval package and identifies the key studies contained within.

Medline

> **Key Points**
> - Secondary resources allow clinicians to efficiently locate relevant primary literature to answer questions.
> - Care must be taken to ensure that the important terms in a question are part of the search strategy used.
> - The proper use of the correct terminology (controlled vocabulary) for each secondary resource will help to retrieve the most relevant search results.
> - A thorough search for relevant primary literature should include accessing multiple secondary sources due to the variation in journal coverage among these resources.

- Produced by US National Library of Medicine.
- Available through various vendors and PubMed. PubMed is the US National Library of Medicine's free gateway to access Medline as well as several other databases.
- Indexes over 5500 biomedical journals from 70 different countries.
- Includes abstracts for more than 75% of the records indexed.
- Medline uses a controlled vocabulary, MeSH, to index topics. MeSH terms are assigned to each citation to describe the main subjects of the journal article. Using MeSH terms when searching Medline ensures that the citations obtained describe the main topics of the search terms entered. This minimizes irrelevant results that may be obtained if general, free-text search terms are used. With the use of free-text terms, citations can be retrieved in which the search term was mentioned in the journal article but was not an important part of the article.
- A MeSH database is available in PubMed to help identify the proper MeSH terms to use.
- MeSH terms are arranged hierarchically by subject category with broader terms listed above more specific terms. If a broad MeSH term is used, Medline will automatically include all of the more specific terms listed beneath it (referred to as "exploding" the term).
- PubMed also includes a comprehensive tutorial. This tutorial helps to explain the many features of PubMed for clinicians inexperienced in using this resource.

Self-assessment questions

Question 1
A physician would like to read clinical trials evaluating the use of ketorolac and meperidine or ketorolac and diclofenac for relief of pain associated with kidney stones. What are the key terms from the question that should be part of a search strategy?

Question 2
How would you combine the above terms in a search of a secondary database?

Question 3
After running the search above, several irrelevant results were returned. Can anything else be done to refine the results better?

Question 4
A patient would like to know if a drug that was inadvertently left out of the refrigerator for several hours could still be taken. Which secondary resource would be most useful in researching this question?

Question 5
A clinician wishes to know if there have been reports of an allergic reaction to insulin and would like to find primary literature discussing insulin allergies. What are the key terms in this question that should be included in a search strategy?

Question 6
How would you combine the above terms in a search of a secondary database?

Question 7
Is there another way in which multiple endings for the word "allergy" could have been searched all at once?

Summary

An efficient search strategy should use a systematic process that includes identifying key terms in the question, selecting appropriate database-specific search terminology for those key terms, and combining the search terms using appropriate Boolean operators. Available secondary resources index a variety of different journals, so multiple secondary resources should be consulted to provide thorough search results. This chapter provided a brief overview of common search techniques to use when accessing secondary resources, as well as important characteristics of secondary resources commonly used by clinicians.

4

Internet resources

Learning objectives

Upon completion of this chapter, you should be able to:

- discuss consumer use of the internet to find medication and health information
- describe problems when using the internet to search for medication- and health-related information
- identify specific types of medication information questions for which it is appropriate to use the internet to find information
- identify criteria that should be used when evaluating medication and health information on the internet
- list useful internet sites that provide high-quality medication and health information to consumers and/or practitioners.

Introduction

Consumers are increasingly taking interest in their healthcare, which includes gaining knowledge about different treatment options available, participating in preventive care programs, or asking more direct questions about the care they are receiving. Their interest may be motivated by the skyrocketing costs associated with healthcare or simply to obtain the best care possible. Regardless of the reason, many consumers are turning to the internet to research information about their medications, health conditions, lab tests, and other aspects associated with their health. A 2010 Harris Poll estimated that 175 million American adults have used the internet to seek health information: 62% of those have looked online for health information in the past month, 32% report they often search for health information online, and 17% of respondents used the internet 10 or more times during the prior month to search for health information. Only 8% of respondents felt the health information they found was unreliable.

However, in contrast to what most consumers believe, the information found on internet websites is not always accurate and complete. The quality and reliability of the information provided can vary among sites, and consumers and

> **Key Point**
>
> Pharmacists and other healthcare practitioners should take a lead role in helping patients to locate and evaluate high-quality, reliable internet-based information, and to apply the information found appropriately to the patient's personal health needs.

practitioners alike should be aware of this when searching the web for health-related information. Variability in quality is a problem because there is no agency/entity that routinely reviews all internet sites. It is relatively easy for anyone to publish a website, and there is no enforcement of any rules related to the content or type of health information provided. For this reason, every website should be evaluated to ensure that the health information contained is reliable and of high quality.

Worked example

Example 4.1

Consider the following excerpt from an actual website:

Bee pollen

> There are 22 basic elements in the human body – enzymes, hormones, vitamins, amino acids, and others – which must be renewed by nutrient intake. No one food contains all of them, except bee pollen. The healing, rejuvenating and disease fighting effects of this total nutrient are hard to believe, yet are fully documented. Aging, digestive upsets, prostate diseases ... and a host of other conditions have been successfully treated by bee pollen. Vigorous good health has been maintained by the millions of people who make bee pollen a staple of their diets.

Needless to say, the claims above would not be taken seriously by most healthcare providers. The website states these claims are fully documented but does not provide any actual documentation. However, a consumer who views this site may believe the claims made and rush to purchase bee pollen. Healthcare practitioners need to maintain an open dialogue with patients about internet-based information so they can dispel any misleading or incorrect information.

Evaluating the quality of internet-based medication and health information

Key Point

If a website uses sensationalized terms or phrases to describe a product or treatment, such as "miracle cure," "wonder drug," "vital ingredient," "scientific breakthrough," or "amazing results," almost invariably the claims will be "too good to be true." This type of language should be viewed with skepticism. Sensationalized terms are used to catch the patient's eye but often simply cover the fact that there is not good information available to substantiate use of that product or service.

In general, websites maintained by governmental agencies (.gov), educational entities (.edu), or professional associations and nonprofit organizations (.org) tend to provide higher-quality health information that is reliable and accurate. Commercial entities (.com) may also provide appropriate information, but since these sites may be selling the products being discussed there is greater risk for biases and inaccuracies to be present.

Several characteristics should be considered when evaluating the health information provided by a website, including the following:

1. Content:
   a. The information should be presented at an appropriate level for the intended audience.
   b. The information should be accurate, complete, and presented in a logical manner that is balanced and neutral.
   c. Evidence of peer review or an editorial review board may lend credibility to the accuracy and completeness of the information (although this will often not be available).
   d. Sensational or overly positive words such as "miracle ingredient," "amazing results," and "vital product" should be avoided.
   e. The text should be free of grammatical and spelling errors.
   f. Opinions should be labeled and represented as such and should not be presented as facts.
   g. Opinions, conclusions, and summaries presented should be consistent with factual information.
   h. Any hyperlinks contained should link to relevant and reputable websites.
2. Currency:
   a. Determining how current the information is should not be based solely on when the website was last updated, although this is important. It should be based on a review of the information presented. The date of last update is sometimes included on websites, but this does not necessarily mean that all of the content was updated at that time.
   b. Review the information to determine if it is current. Checking the references listed, or comparing the information with any recent work in the field, may help evaluate if the information is up to date.
3. Authors/source:
   a. The authors or source of the information should be identified. If the information is from an author, he/she should have recognized credentials and an appropriate background for the type of information provided. If the information is from another source (organization), it should be reputable and the information presented should be related to the source's function (e.g., the American Cancer Society providing chemotherapy information).
   b. Contact information for the authors/source should be provided to enable a reader to ask questions or request additional information.
   c. The purpose of the website should primarily be for education and not for product promotion.
   d. The website should be free of bias and authors should not be financially affiliated with the products or services discussed. Any advertisements present should not be related to author-affiliated products or services or to any manufacturer whose products or services are discussed.

4. References/documentation:
   a. References should be cited completely so the information can be verified, and they should be from reputable sources.
   b. They should be scientifically sound and, if related to therapy, should refer to clinical trials in humans.
5. Site design and organization:
   a. The design should allow for easy navigation through the material. If the information is lengthy, an index or topic outline should be present to allow sections of interest to be readily accessed.
   b. Overly flashy graphics should be avoided since they may be used to distract from inaccurate or incomplete information.
   c. There should be a prominent disclaimer that states the health information is not intended as medical advice or to replace a healthcare provider.
   d. The site's privacy policy should be readily available and easily accessed. The site should disclose if they collect or track any personal information. If the site does collect or track that information, the purpose for doing so should also be transparent.

Key Point

Use clinical judgment when assessing references. If references are not used but the source of information is from a reputable expert resource, such as a National Institutes of Health agency, the information presented would be expected to be credible. However, if a site providing information about a neurological disease was authored by a general practitioner, a lack of references would be more concerning.

See Table 4.1 for a list of questions to ask when evaluating internet-based health information. For additional information, refer to the Medical Library Association's website (http://www.mlanet.org/resources/userguide.html) and the National Library of Medicine's MedlinePlus website (http://www.nlm.nih.gov/medlineplus/healthywebsurfing.html).

The HONcode is another tool patients and practitioners can use to assess the quality of an internet site. The HONcode is the code of conduct of the Health On the Net Foundation. This organization provides guidelines for site developers to use to create websites that provide high-quality, objective, and transparent medical information tailored to the needs of the audience. If a site meets the criteria for the code of conduct, there will be a symbol with the words HONcode placed at the bottom of the web page. This is a voluntary program that each site must apply for. Thus, the lack of the HONcode symbol on a web page cannot be relied upon to indicate that the information provided is of poor quality. In addition, although the HONcode defines a set of rules for ethical standards in the presentation of information, the Health On the Net Foundation does not evaluate the accuracy or completeness of the information presented.

Table 4.1 Questions to ask when evaluating internet-based medication and health information

Criteria	Questions to ask
Content	1. Is the patient the intended audience for the information? If yes, is the information written at the appropriate level for patients? 2. Are proper spelling and grammar used throughout? 3. Is the information in the site well organized and presented in an easy-to-follow manner? 4. Is the disease/medical condition or drug/therapy information complete (e.g., covers key areas, free of notable or otherwise significant omissions)? 5. Is the information properly balanced and neutral in its approach to the topic (i.e., does it provide both positive and negative data, pros/cons, risks/benefits in a factual manner without undue emphasis on one or the other)? 6. Are sensational or extremely positive words avoided? 7. Are any opinions, summaries, or conclusions expressed? If yes, are they based on (consistent with) facts? Are opinions stated as such? 8. Does the information appear to be accurate and valid (e.g., no errors identified, clinically unsubstantiated claims for treating, preventing, or curing diseases are absent, areas requiring further research are clearly stated, information is not taken out of context)? 9. Is there any evidence of peer, editorial, or other review of the site's content? 10. Does the site contain hyperlinks to other websites? If yes, are the links relevant and reputable? 11. Is the quality of information at the site at least as good as (or better than) that from other sources?
Currency	1. Does the information appear current (e.g., based on a copyright or update date, the references listed, or knowledge of recent work in the field)?
Authors or source of information	1. Are the authors or organization(s) identified? 2. For information authored by individuals: Do they have a recognized professional degree (e.g., BS, MS, PharmD, MD, PhD, DDS), and are their credentials (e.g., title, position, experience) appropriate for the type/level of information provided? 3. For information authored by an organization(s): Is it a reputable entity (e.g., government agency, university, professional association)? Is the information provided related to its function (e.g., American Heart Association providing hypertension information)? 4. Do the authors or organization(s) provide ways to contact them to ask questions or request additional information? 5. Is there an apparent absence of bias or conflicts of interest by the authors or organization(s) producing the site (e.g., no financial ties to products/services discussed, authors not consultants to or employees of the product/supplier of products/services discussed)? 6. Does the purpose of the site appear to be primarily for education (e.g., non-commercial site, primary purpose not to sell products/services of authors/site developers)? 7. Are advertisements absent or, if present, are they unrelated to books, products, or services produced or supplied by the site producer?

(continues)

Table 4.1 *(continued)*

Criteria	Questions to ask
References	1. For the information, are the sources used for the information provided or cited? References may include other internet sites. 2. Are references cited completely so that the information can be verified? 3. Is the information or references from standard medical/healthcare journals, medical textbooks, or other reputable sources (e.g., government agency publications or their websites, statements/information sheets prepared by professional medical/healthcare organizations or societies)? 4. Do references appear to be scientifically sound (e.g., randomized controlled trials, meta-analyses, fact sheets or statements prepared by government agencies)? 5. For information related to the possible use of a medication/therapy by patients, do references cited appear to describe clinical trials in humans (as opposed to a preponderance of animal studies or *in vitro* work)? 6. If additional recommended reading materials or resources are provided, do they appear to be scientifically sound?
Site design and organization	1. Is the information designed in a manner that makes it easy to navigate through the material contained? 2. If the material is lengthy, does it have a map or topic index readily available to assist a patient in finding specific information? 3. Is there a prominent (e.g., easily visible without the need to access a link) disclaimer describing the limitations of the information or a statement emphasizing that the content is not intended as medical advice or to replace a healthcare provider?
Privacy policy	1. Is the website's privacy policy readily available? 2. Does the website collect any information about visitors to the site? If yes, is it clear how that information is used?

Internet resources

A general internet search can be useful and appropriate when looking for certain types of medication- or health-related information. For example, it would be difficult to find information about very recent topics using other resources. Patients will often ask medication-related questions based on a story they heard on the local news or read in a newspaper. Searching the various television network and general news wire internet sites is a good way to determine exactly what the patient heard and the source of this information (e.g., recently published clinical study). Internet searches can also be very useful for researching the ingredients of marketed over-the-counter products, or multi-ingredient dietary and herbal products.

Useful internet sites for clinicians

There are several reliable websites that clinicians can use to find medical and health-related information of a variety of types. These websites and some of the information they provide that might be particularly helpful for practitioners are summarized below. For additional information about the resources offered by the National Library of Medicine, refer to the National Library of Medicine's website (http://wwwcf2.nlm.nih.gov/nlm_eresources/eresources/search_database.cfm).

Government-sponsored sites

AIDSinfo (www.aidsinfo.nih.gov)
- Includes guidelines for HIV/AIDS treatment, transmission, management of complications, postexposure prophylaxis
- Includes information for Food and Drug Administration (FDA)-approved and investigational drugs used to treat AIDS
- Provides information for clinical trials and education materials
- Health topics section links to information and resources for HIV/AIDS.

Agency for Health Care Research and Quality – AHRQ (www.ahrq.gov)
- Evidence-based practice section contains documents that review literature on select topics
- Effective healthcare section includes guides and resources for different treatment options
- Preventive services section includes screening guidelines, health promotion information, wellness practices
- A link to the National Guideline Clearinghouse is provided in addition to some additional guidelines.

Centers for Disease Control and Prevention – CDC (www.cdc.gov)
- An excellent source of international travel information (vaccines, disease prophylaxis, food and water precautions), vaccines and immunizations
- Includes *Morbidity and Mortality Weekly Reports* (MMWR) that discuss health statistics, vaccine recommendations, therapy of sexually transmitted diseases; the journal *Emerging Infectious Diseases*; and *Preventing Chronic Diseases*
- Diseases and conditions section contains an alphabetical index of numerous disease topics with information about those conditions
- Staying healthy section contains a wide array of information for preventive care

- Includes sections with information about environmental health, emergency preparedness and response, data and statistics on a variety of topics, and global health.

ClinicalTrials.gov (www.clinicaltrial.gov)

- Developed by the National Institutes of Health through the National Library of Medicine in collaboration with the FDA
- Provides up-to-date information for locating clinical trials (ongoing/recently completed) for a wide range of diseases or conditions
- Contains over 125 000 trials sponsored by the National Institutes of Health, other federal agencies, nonprofit organizations, and private industry
- Information includes disease or condition and experimental treatments studied, study design, purpose, sponsor, recruiting status, inclusion/exclusion criteria, location, and contact information.

Dietary Supplements Labels Database (http://dietarysupplements.nlm.nih.gov/dietary/)

- Product label information for over 7000 brands of dietary supplements
- Includes product ingredients and manufacturer's information
- Allows a comparison of the label ingredients among brands
- Database can be searched by brand names, uses noted on product labels, specific active ingredients, and manufacturers.

Directory of Information Resources Online – DIRLINE (http://dirline.nlm.nih.gov)

- Directs users to a variety of information sources
- Includes organizations, agencies, centers, societies, support groups, and academic/research institutions
- Includes names, addresses, phone numbers, services, and publications.

US Food and Drug Administration – FDA (www.fda.gov)

- Contains information about human and veterinary drugs, vaccines, and food
- Contains the electronic "Orange Book" that will list the products that have therapeutic equivalence to other specified products
- Provides FDA news items about labeling changes and new drug approvals
- Drugs@FDA database provides drug-related FDA materials (e.g., approvals, labeling)
- Drug safety and availability section provides medication guides, drug shortages, drug safety communications, drug recalls, MedWatch adverse event reporting program, and other safety announcements

- Animal and Veterinary tab on FDA home page will provide information about approved veterinary products, including medications and food/feed. Recalls and other safety information are also provided.

Healthfinder (www.healthfinder.gov)

- Developed by the US Department of Health and Human Services
- Directs users to internet sites with high-quality information
- Uses specific guidelines to evaluate information included. Information is generally from government agencies, nonprofit organizations, and universities
- Quick guide to healthy-living section contains preventive care information
- Personal health tools section contains online checkups, activity and menu planner to help count calories, and other health calculators
- Health A to Z section gives links to information on over 1600 health-related topics
- Health news section provides updates for recent health-related news information
- Find Services and Information section includes searches for healthcare providers, home healthcare, public libraries, and health organizations.

Health Hotlines (http://healthhotlines.nlm.nih.gov/)

- Derived from DIRLINE
- Provides toll-free phone numbers for over 14 000 biomedical resources and over 300 organizations
- Also includes information about services and publications available in Spanish.

MedlinePlus (www.nlm.nih.gov/medlineplus/)

- Developed by the National Library of Medicine (NLM)
- Excellent patient information resource but also useful for health professionals
- Provides information about diseases, conditions, and wellness topics
- Includes easy-to-read information for select topics
- Health topics section provides links to selected internet sites that provide a variety of high-quality information
- Drugs and supplements section provides information about medications that is derived from AHFS Consumer Medication Information (produced by the American Society of Health-System Pharmacists). The herbs and supplement information is derived from either the Natural Medicines Comprehensive Database or the National Center for Complementary and Alternative Medicine of the National Institutes of Health

- Also contains a medical dictionary and medical encyclopedia, a directory to find US healthcare providers, libraries, hospitals, and healthcare services, and a list of health organizations.

National Cancer Institute – NCI (www.cancer.gov)
- Provides information on types of cancers, treatments, prevention, genetics, causes, screening and testing, coping with cancer, support and resources, and basic statistics
- *NCI Drug Dictionary* contains definitions and synonyms for drugs used to treat cancer. Each entry also has a link to check for clinical trials in the PDQ cancer clinical trials registry
- PDQ cancer clinical trials registry is NCI's listing of over 10 000 cancer clinical trials accepting participants.

National Center for Complementary and Alternative Medicine – NCCAM (http://nccam.nih.gov)
- Provides useful information about a variety of complementary and alternative medicines (CAM), including herbals, supplements, and other therapies
- Includes a listing of clinical trials involving CAMs
- Includes monographs and practice guidelines for CAMs.

Toxnet (http://toxnet.nlm.nih.gov/)
- A collection of several different databases related to toxicity/toxicology
- Toxline contains a bibliography database that focuses on biochemical, pharmacological, physiological, and toxicological effects of drugs and other chemicals
- LactMed is a peer-reviewed and fully referenced database of drugs and other chemicals to which breastfeeding mothers may be exposed. It includes maternal and infant drug levels, adverse effects in infants, effects on lactation, and alternate drugs to consider
- Other databases include Hazardous Substances Data Bank, Household Products Database, Developmental and Reproductive Toxicology Database.

Nongovernment sites

Doctor's Guide to Medical and Other News (www.docguide.com)
- Free registration; can customize desired information
- Provides weekly e-mail updates, e-newsletters on a variety of health topics in specialty and subspecialty areas
- Good source of health-related news, literature abstracts.

MD Consult (www.mdconsult.com/php/346870355-16/home.html)
- Provides information covering 31 medical specialties
- Provides access to full text of 50 medical references and full-text articles from 80 medical journals and The Clinics of North America
- Includes drug information provided by Gold Standard (publisher of *Clinical Pharmacology*)
- Also includes over 1000 practice guidelines and 15 000 patient handouts.

Medscape (www.medscape.com)
- Free registration, can customize the information by specialty area
- Provides health-related news
- Provides condition-specific information for a variety of topics
- Provides brief overview information for drugs, diseases, and procedures
- Can receive free e-newsletters and news clips via e-mail.

Natural Medicines Comprehensive Database (http://naturaldatabase.therapeuticresearch.com/home.aspx?cs=&s=ND)
- See Chapter 2 for a description of the Natural Medicines Comprehensive Database.

Natural Standard (http://www.naturalstandard.com/)
- See Chapter 2 for a description of the Natural Standard databases.

RxList (www.rxlist.com)
- Contains drug package inserts
- Includes patient drug information
- Contains supplement monographs derived from the Natural Medicines Comprehensive Database Consumer Version.

Product-testing sites

ConsumerLab (www.consumerlab.com)
- Independent testing lab that verifies identity, potency, purity, bioavailability, and batch consistency for health and nutritional products
- Requires a subscription but is relatively inexpensive.

USP Verified (www.usp.org/USPVerified)
- US Pharmacopeia testing program that is voluntary by manufacturers of dietary supplements
- If the label has the USP Verified seal, the supplement has been tested to assure integrity, purity, dissolution, and safe and appropriate manufacturing; however, the USP does not evaluate efficacy.

News agency sites (all of the following sites allow the practitioner to search for health-related news stories)

- Associated Press (www.ap.org)
- PR Newswire (www.newswire.com)
- Reuters Health News (www.reutershealth.com).

Mobile resources

> **Key Point**
>
> Mobile resources are useful for finding quick information, but should not be a substitute for using other, more thorough resources. Many of the mobile applications are streamlined versions of the full databases and may not include all of the features found on the full database versions. Use the mobile applications when access to other resources is impractical, but do not rely solely on mobile resources for all questions received.

It is ideal to use a combination of the resources discussed in the previous chapters when researching a medication information question. However, there are times when a busy clinician needs to find information quickly and in a setting where the use of the full armament of available resources is not possible. For example, a clinician rounding with a medical team may be asked to provide information quickly about a specific medication to be used for a patient. In these circumstances, the use of a mobile device (smartphone, tablet, or PDA) can greatly aid the clinician. There are several applications available in a variety of platforms that can be used with mobile devices that provide readily accessible information about a variety of medications and other health-related topics. Commonly used applications for mobile devices include Medscape, MobileMicromedex, Epocrates, MobilePDR, *Clinical Pharmacology* OnHand, and LexiComp. See Chapter 2 for a description of *Micromedex*, PDR, *Clinical Pharmacology*, and LexiComp. Medscape is discussed previously in this chapter.

Epocrates

- Free for a basic subscription, more advanced features available for a fee
- Includes monographs with basic medication information, a drug identification tool, a drug interaction tool, and various clinical calculators
- A subscription fee allows access to a disease database, a medical dictionary, alternative medicine information, and an infectious disease guide
- Includes updates that are often sponsored by drug manufacturers and these should be screened for bias by clinicians using Epocrates.

Key Point

- Millions of patients are using the internet to find health-related information. Most of these individuals do not believe the information they find is unreliable, even though many of the websites contain information that is not accurate or is misleading.
- There is no regulatory agency that routinely evaluates the quality or reliability of health information posted on internet websites.
- Every practitioner should have a working knowledge of the criteria used to evaluate the quality of internet-based information, and they should help patients identify reliable websites and assist patients in understanding how the information may, or may not, apply to a patient's unique situation.
- Websites developed and maintained by governmental agencies, educational entities, and professional or nonprofit organizations are more likely to contain high-quality, reliable health information than commercial sites.
- The HONcode symbol can give some assurance that a website maintains high standards but the information still needs to be evaluated for accuracy and completeness.
- Several medical or health-related websites exist that can provide clinicians with useful information to answer questions that arise during practice.

Self-assessment questions

Question 1
A patient tells you she found some great information on the internet about her diabetes. She states she knows she found complete and accurate information because the site had the HONcode symbol. Is the patient correct?

Question 2
In addition to a copyright and date when last updated that might be listed on some websites, what else should be considered when evaluating the currency of the information provided?

Question 3
For what types of medical topics might a general internet search be especially useful as an information resource?

Question 4
Name two websites that would be especially useful to recommend to patients as sources for locating high-quality internet health-related information.

Question 5
A patient will be traveling to Brazil in 2 months. He wants to know if he will need any immunizations or will need to take any medications in preparation for this trip. Where should a clinician look to find this information for the patient?

Question 6
A patient brings to the pharmacy a prescription for Unithroid. The pharmacy usually stocks Synthroid brand and a generic levothyroxine that can be substituted for Synthroid. You are not sure if the generic levothyroxine also has therapeutic equivalence with Unithroid. Which website would be expected to provide this information?

Question 7
A physician tells you about a patient who was recently diagnosed with metastatic prostate cancer. The physician remembers reading something about a clinical trial of a new drug that might help this patient, but can't remember any of the specifics. Is there a way to find this information?

Question 8
A patient approaches a pharmacy counter with two different bottles of fish oil capsules. One bottle has a USP Verified seal on the label while the other does not. The patient asks you for a recommendation of which fish oil product to purchase. Both products have about the same cost. Which product would be best to recommend to the patient?

Summary

Patients are increasingly using the internet to search for health-related information. The internet can be a valuable source for many types of medication and health information, but it does have problems. Not all websites are reputable and misinformation, inaccurate information, or incomplete information can lead to negative consequences for the patient. Healthcare providers need to take an active role in helping patients find high-quality internet information, to apply that information to meet patients' needs, and to help patients evaluate the health information found. This chapter provided an overview of the criteria used to evaluate internet health information sites, and it briefly described several websites that could be particularly useful resources for healthcare professionals.

5

Evaluating clinical studies – step 1: what type of study is it?

Learning objectives

Upon completion of this chapter, you should be able to:

- describe the main types of study designs
- describe the features, advantages and disadvantages of the observational study designs
- explain why the overall study design is important when evaluating studies and applying their findings to practice.

Introduction

There are two broad types of published studies: descriptive (simply recording information from observing patients) and explanatory (using group comparisons as the basis for determining whether an exposure/treatment might cause or affect a condition or outcome). Descriptive studies include case reports (reporting observations in one or a small number of individual patients) and case series (reporting observations from a small group or series of patients). Since descriptive studies are generally not considered to be "studies" and are normally referred to as "reports," we will focus on explanatory studies (consisting of experimental and observational studies).

Experimental studies

Experimental studies (controlled or noncontrolled) involve *actual intervention* by the investigators (i.e., subjects are assigned and given the treatments by investigators). Controlled experimental designs are best (i.e., "gold standard") since they use a treatment group(s) and a control (comparison) group(s). Types of control groups include placebo, active (use of another treatment with established efficacy for the condition studied), no treatment, or historical (comparison with a treatment previously studied; not commonly used and only when it is not possible to use a different type of control). The control group helps account for factors (other than the treatment) that might affect the study results. Investigators compare the effects seen in the control patients with those in the treatment patients to determine if there is a difference between them.

One unique type of experimental study has been called the "*n*-of-1" or single-subject research design. With this study, the researcher, often a primary care practitioner, identifies a specific patient to study. The researcher conducts a baseline assessment of the patient's condition, followed by therapy initiation. During/after therapy, the researcher measures changes in the condition. The researcher might decide, after stopping therapy and reassessing baseline measures, to repeat the therapy to determine if the same effects are again observed. If there is another therapy to study, after stopping the first therapy the researcher would reassess the baseline measures and then initiate the next treatment, repeating the same measurements during/after therapy. While the *n*-of-1 trial can be beneficial for studying individuals with rare conditions or those requiring unique, individualized therapy, its disadvantages do not allow it to replace other experimental designs. Readers should refer to a review by Janosky (2005) for more information about *n*-of-1 studies.

> **Key Point**
>
> **Disadvantages of *n*-of-1 studies**
> - Inability to generalize results to others
> - Difficult to impossible to perform statistical analyses
> - Difficult to validate findings.

Since controlled experimental studies are critically important for determining therapy efficacy and in making clinical therapeutic decisions, this type of study will be the focus of subsequent chapters.

Observational studies

Case-control, cohort, and cross-sectional studies are *observational* designs, meaning the treatment(s) taken or other exposures studied were *not given* by the study investigators. Although the controlled experimental study design is best, observational designs are generally used when it is not possible, feasible (e.g., for rare conditions or those that require a long time to develop), or ethical to use an experimental design. In these situations, they can provide very helpful information.

> **Key Points**
>
> **Example – coffee consumption and pancreatic cancer**
>
> Investigators are interested in studying whether coffee intake is associated with an increased risk of pancreatic cancer development. They suspect that coffee might be a risk factor for pancreatic cancer.
>
> **Would it be appropriate for the investigators to use an experimental design to test their hypothesis?**
>
> No. It is unethical for investigators to administer a therapy (coffee) to study subjects for the primary purpose of determining if they are more likely to develop an adverse

outcome (pancreatic cancer, usually fatal), without any possible offsetting advantages. Since cancer generally takes a long time to develop, an experimental study would also not be practical. An observational study design enrolling subjects already drinking coffee or diagnosed with pancreatic cancer would be appropriate here.

Table 5.1 provides a summary of each of the observational designs. The *case-control* design is used to determine the possible factors (e.g., exposures, drugs) influencing or causing an event or outcome. It is always *retrospective* (looking backward). Why? This design begins with patients who already have the event or outcome (cases) and another group of similar patients who lack the event or outcome (controls). The investigators need to look back in time in order to compare drug use or the extent of exposure in both groups *prior to* when they developed the outcome. If the cases are found to have significantly greater drug use or extent of exposure than the controls, a possible association exists between the drug/exposure and outcome development.

Key Points

Prospective versus retrospective cohort studies

A cohort study can be *prospective* (*concurrent*) or *retrospective* (*nonconcurrent, historical*) in nature. The basic design of each is the same: (1) first identify groups (cohorts) with and without the drug use/exposures of interest – no one has the outcome at the start; (2) follow the groups forward over time and measure differences in outcome development.

The nonconcurrent or retrospective design differs from the prospective cohort study in that *all* information (drug use/exposures and outcomes) is obtained from already existing medical records or databases. The start of a nonconcurrent cohort study occurs at a designated point in the past. The investigators initially select the cohorts for inclusion in either the study or control groups *with no knowledge* of whether or not the outcome later develops. Once all subjects are included the investigators examine the existing data, going forward in time from the starting point, to determine whether or not the subjects in each group developed the outcome of interest.

Which cohort design, prospective or retrospective, is strongest?

The prospective concurrent design is best because it is less subject to bias and inaccuracies. The nonconcurrent or retrospective design is dependent upon existing records or databases that might be incomplete or incorrect.

The *cohort* design follows a study "cohort" (a group of individuals/subjects who share a common characteristic) over time to determine if a drug or other exposure will lead to the development of an outcome of interest. Unlike the case-control design, the subjects in a cohort study do not have the outcome at the start of the study. Instead, investigators identify subjects who are taking the drug or have the exposure of interest (study subjects), as well as similar subjects who are not taking the drug or who lack the exposure (control/comparison subjects). The investigators then follow the subjects

Table 5.1 Observational study features, advantages, and disadvantages

Study type	Key features	Advantages (A) and disadvantages (D)
Case-control	**At start** Select: Cases: patients who already have condition or outcome being studied Controls: patients who do not have the condition or outcome being studied but are otherwise similar to cases **Compare in the groups**: Extent of exposure to drugs or other factors thought to affect development of the condition or outcome: done *retrospectively* (looking back in past) through use of surveys, interviews, medical records, or medical databases **Analyze and determine**: Were cases more likely than controls to have been exposed to drugs or other possible causes of the condition/outcome?	A: Good for studying possible causes of adverse events or negative outcomes, especially rare/infrequent outcomes or those that take a long time to develop; faster, less expensive than prospective study designs D: Possible selection bias (are cases truly comparable to controls?); recall bias (patients' memories of an exposure might be inaccurate); interviewer/observer bias (interviewer might slant data collection if aware of who is a case or control); records or databases could be inaccurate or incomplete
Cohort (follow-up)	**At start** Select: Study subjects: subjects already taking certain drug(s) or who have an exposure(s) that might affect the outcome of interest Control/comparison subjects: subjects who are not taking the drug(s) or who do not have the exposure, but who are otherwise similar to study subjects **Compare in the groups**: Extent to which subjects in each group develop the condition/outcome: done by following subjects forward over time (*prospectively*, if starting point in present time and data about outcome development obtained in future; *retrospectively*, if complete medical records exist over an extended time that allow the starting point (exposure or no exposure) to be in the past, with data about outcome development obtained by looking through subsequent medical records) **Analyze and determine**: Were the study subjects more likely than the comparison subjects to develop the condition or outcome?	A: Good for studying possible causes of adverse events or outcomes especially if they occur relatively commonly, or how/if certain exposures or characteristics might affect later outcome development D: Possible selection bias (are study subjects truly comparable to control/comparison subjects?); with prospective design: can take long time to complete, subject drop-out (loss to follow-up) could occur, potential expense; retrospective design shares disadvantages of case-control design

Table 5.1 (continued)		
Study type	Key features	Advantages (A) and disadvantages (D)
Cross-sectional (prevalence, survey)	**At start:** Subjects identified who represent a population of interest; they might/might not have the exposure(s) or the outcome(s) of interest **Compare in group:** Presence and absence of both exposures and the condition or outcome of interest: determined at same time, resulting in "snapshot" or "cross-section" of the exposures and outcomes present at one given time (done through use of surveys, questionnaires, or examination of databases or patient records) **Analyze and determine:** What are the factors or exposures associated with certain conditions or outcomes, and what is their prevalence?	A: Can identify possible risk factors for or potential causes of a disease or condition, the prevalence of a disease or condition at a specific point in time, or if beliefs or practices affect health or behavioral outcomes; relatively quick, inexpensive D: Cannot determine whether the "cause" (characteristics or exposures) actually preceded or resulted in the "effect" (outcomes); group identified for study inclusion might not adequately represent desired population of interest; subject self-reporting in surveys or questionnaires might not be accurate or nonresponse might be a problem

in both groups (through scheduled visits, by examining medical records) over a certain period of time to compare the extent to which they develop the outcome. If significantly more subjects in the study group develop the outcome compared to the control subjects, it is concluded that the drug or exposure might contribute to outcome development.

In a *cross-sectional* design, the study sample is selected from a targeted population of interest and information about both the extent of drug use/other exposures as well as the presence of the outcome is obtained from the sample at the same time. Thus, the cross-sectional study provides a "cross-section" snapshot of the prevalence or existence of specific conditions, characteristics, and outcomes at *one point in time*. The investigators obtain all the exposure and outcome information from the study sample through the use of questionnaires or surveys. The data from subjects within the sample are compared and analyzed based on the presence or absence of these factors. Since a cross-sectional study collects data about past exposures or drug use from subjects' recollections or records, it is subject to similar limitations as the case-control study. The cross-sectional study also lacks a separate control/comparison group.

Please refer to Abate (2012) for further details about the types of studies and their advantages and disadvantages.

Worked example

Example 5.1

Investigators wish to study if aspirin might lead to Reye's syndrome development in small children who take aspirin for a viral illness. Children with a viral illness who later developed Reye's syndrome were identified, along with other children with a viral illness who did not develop Reye's syndrome. The use of aspirin during the viral illness, including dose and duration of therapy, was analyzed and compared in both groups through interviews with parents.

What study design was used?
Case-control. The children with Reye's syndrome (condition/outcome present) were the cases, and the children without Reye's syndrome were the controls.

What was compared/analyzed?
Use of aspirin in both groups. Since the exposure (in this case, aspirin use) must *precede* development of the condition/outcome (Reye's syndrome), information about the exposure is obtained by looking in the past (retrospective). Greater aspirin use in the cases might indicate a link between aspirin use and development of Reye's syndrome. However, even if aspirin use was much greater in the cases, this study design cannot *prove* that aspirin causes Reye's syndrome.

Worked example

Example 5.2

Previous studies in adults have indicated a possible link between low vitamin D levels and depression symptoms. Since children can also develop depression, investigators studied whether low vitamin D levels might be associated with depression in children. Several hundred children were identified and their vitamin D concentrations were measured. The children were then divided into groups based upon their vitamin D level: low, normal, or high. The children were followed over the next 3–4 years and were periodically tested to determine if they developed symptoms of depression.

What study design was used?
Cohort. The children were initially divided into groups based upon the extent to which they had the exposure/factor of interest (vitamin D level); the study subjects had low vitamin D levels and children in the comparison groups had normal or high levels. There was no therapy intervention. The investigators did not administer vitamin D to the children but rather measured existing concentrations, so the design was not experimental.

What was compared/analyzed?
Development of depressive symptoms (condition/outcome) in all children. Children were followed prospectively over time to determine the extent to which depression occurred. Greater depressive symptoms in children with low vitamin D levels compared to those with normal to high levels could indicate a link between vitamin D deficiency and depression. However, even if depression developed to a much larger degree in children with low vitamin D concentrations, this study design cannot *prove* that a vitamin D deficiency causes depression.

A summary of key points and how to apply the information from this chapter to practice follow.

Key Points

- In general, the order of the study designs from strongest (best) to weakest (most limitations/disadvantages) is: (1) controlled experimental; (2) prospective cohort; (3) case-control/cross-sectional/retrospective cohort.
- Due to their possible disadvantages, observational studies (case-control, cohort, cross-sectional) *cannot prove* that a drug or exposure caused a certain outcome; only well-designed controlled experimental studies can do this.
- Observational studies can still provide very useful information when it is not possible, feasible, or ethical to conduct an experimental study (for example, to study whether a drug or other exposure might cause an adverse outcome or to study the factors that might predispose to development of a rare or infrequently occurring condition).

How to apply to practice

- If a news report claims that a drug or other exposure causes a certain adverse outcome based upon findings from an observational study, this might not be accurate. Further investigation is generally needed for confirmation.
- For the weakest study designs (case-control, cross-sectional, retrospective cohort), confirmation of their results by further study is even more important.

Self-assessment questions

Question 1

A study was conducted to determine whether daily users of nonsteroidal anti-inflammatory drugs (NSAIDs) were at lower risk of developing benign prostatic hyperplasia (BPH) than nondaily NSAID users. The medical records from a large health center were used to identify men who were either daily (536 men) or nondaily (659 men) NSAID users. The men then received twice-yearly examinations for the next 5 years for signs and symptoms of BPH. After adjusting for age differences between groups, greater daily NSAID use was associated with less BPH development compared to nondaily NSAID use. The authors concluded that regular NSAID use might prevent or delay the development of BPH.

Which type of design was used in this study? Can this study be used to prove that daily NSAIDs can decrease the risk of BPH development?

Question 2

Investigators wish to conduct a study with the following objective: to determine whether statins used to reduce cholesterol might increase the likelihood of developing type 2 diabetes. Briefly describe how this study could be conducted using: (1) a prospective cohort design and (2) a case-control design.

Question 3

A study was performed to determine the effects of garlic powder tablets on blood glucose levels and plasma lipids in patients with type 2 diabetes. Fifty-six type 2 diabetes patients were randomized to receive either two garlic tablets BID or a placebo control for 4 weeks.

Fasting blood glucose was measured daily, and plasma cholesterol and triglycerides were measured at baseline and after 2 and 4 weeks. At the end of 4 weeks, blood glucose and cholesterol levels were found to be significantly reduced in patients receiving the garlic tablets compared to placebo. It was concluded that garlic tablets might be a useful supplement to reduce cardiovascular risk in diabetic patients. Was this an experimental or observational study? Explain.

Question 4

Acetaminophen is often responsible for poisonings and is a leading cause of acute liver failure in the USA. The investigators wanted to study the extent to which adults are knowledgeable about acetaminophen and its potential toxicity and whether this knowledge increased their likelihood of recognizing available acetaminophen-containing medications. Subjects at least 19 years of age who were being seen for a variety of reasons in a large outpatient clinic in Boston were given a survey to assess their knowledge about acetaminophen dosing and toxicity, and whether they could identify commonly used nonprescription combination drug products that contained acetaminophen. Results showed that many patients had difficulty recognizing acetaminophen-containing products by name and were lacking knowledge of acetaminophen dosing and toxicity. No association was found between having greater knowledge of acetaminophen dosing and toxicity and the ability to recognize products that contained acetaminophen.

What type of design was used in this study? Is this considered a strong study design?

Question 5

Which of the following studies is *least* likely to be affected by selection bias: a case-control, cohort, or a controlled experimental study?

Summary

It is important to know the overall design used in a study since each design has inherent strengths and limitations. This chapter reviewed the structure of the observational and experimental study designs along with their advantages and limitations in practice.

6

Evaluating clinical studies – step 2: the journal, authors, and study purpose

Learning objectives

Upon completion of this chapter, you should be able to:
- explain the purpose of a journal's editorial board and use of peer review
- describe the peer review process
- identify potential conflicts of interest for a study's investigators and how they might impact a published study
- discuss the types of information that should be included in a study's introduction
- describe the types of hypotheses and their importance when analyzing a study's findings.

Introduction

There are thousands of medical journals published worldwide; Medline indexed over 5500 journals as of late 2011. These journals and the research studies they publish can vary considerably in quality. Readers should consider a number of factors related to journals, authors/investigators, and the purpose of the research when evaluating published studies. Editorial boards and peer review are two methods important for helping to ensure the overall quality of a journal and its studies. A study's investigators might have potential conflicts of interest that could compromise the objectivity and quality of their work, and the journal should state these when publishing the study. As readers, after considering overall journal quality and potential conflicts of interest, we should examine a published study's objectives and related hypotheses to determine if the design and methods used were sufficient to fulfill the study's intended purpose. This chapter will review important considerations pertaining to journals, investigators, and a study's purpose.

Journal and investigator considerations

Journals should have an *editorial board* to help assure the quality of the studies they publish. An editorial board consists of individuals with expertise

> **Key Points**
>
> **How can one tell if a published study was peer-reviewed?**
>
> There are several ways to determine if peer review was used for a study:
>
> - Look up the journal online (through its website) and read the journal description and the instructions for authors/author guidelines section. The journal will usually state whether they use peer review in these areas and whether all the articles it publishes are peer-reviewed. Alternatively, checking the first pages of print journals may provide information about peer review or refer the reader to where the author guidelines can be found to determine if peer review is used.
> - Most published studies will include the date the manuscript was received by the journal and a date of acceptance by the journal (usually found at the top or bottom of the first page). If the time period between the manuscript's receipt and its acceptance is long enough to allow for peer review (usually 2 months or more), or if a date when the manuscript was revised is included, this can also indicate if peer review was used.

in a journal's area(s) of focus. The board members assist in setting the direction of the journal and in making decisions about the acceptability of manuscripts (research or review papers) submitted by authors/investigators to that journal for publication consideration. The editor(s) will generally read a manuscript first to ensure that it is consistent with the journal's goals and readership. Manuscripts that do not "fit" with the journal or are poorly written or designed would likely be rejected for publication by the editors. For manuscripts that appear reasonable to publish, ideally the editor's decisions about publication will consider the comments and recommendations obtained from peer reviewers.

Peer review involves the journal sending a submitted manuscript to a small number of outside individuals ("peers") who have expertise in that subject area. The peers provide expert review, critique, suggestions for revision, and a recommendation about whether or not to publish. Although the best approach to use for the peer review process (e.g., whether or not reviewers should be "blinded" to the authors' identities) has not been established, peer review is generally considered to be important for improving the quality of published studies. For additional information, refer to an editorial by Balistreri (2007) that provides a brief overview of peer review and important considerations with its use.

Potential conflicts of interest

The investigators should ideally conduct their study in a manner that is free from bias or other factors that might influence or affect their objective judgment. But, sometimes investigators have relationships or other ties, referred to as potential *conflicts of interest* or *competing interests*, that could alter their objectivity. Competing interests could influence the manner in which the investigators conducted the study, reported the results, or interpreted the findings. Potential conflicts of interest can be either personal or financial in nature, although financial conflicts (e.g., ties to a company funding the study) are easier to identify and thus tend to be focused upon.

Key Points

Potential conflicts of interest

Some potential conflicts of interest for a study's investigators/authors include:

- receiving study funding from the manufacturer of one or more of the drugs being investigated
- serving as a consultant for, or on the board of, directors of the pharmaceutical manufacturer of one or more of the drugs being investigated
- obtaining honoraria from or holding stock in the pharmaceutical manufacturer of one or more of the drugs being investigated
- being employed by the pharmaceutical manufacturer of one or more of the drugs studied
- having a personal relationship with, or serving as a representative of, an organization that has a potential conflict of interest related to one or more of the drugs being studied.

Note: Many times pharmaceutical manufacturers will only provide the drugs and placebos used in the study with no other involvement. This is not considered a potential conflict of interest.

Journals should ask authors of submitted manuscripts to disclose any potential conflicts of interest and should clearly state these disclosures when the study is published. Readers should check for the presence of possible competing interests and, if present, consider whether they appeared to affect the study's design, methods, or reporting or interpretation of the results. The presence of potential conflicts of interest does *not* automatically invalidate a study or indicate that the study has major problems; rather, it means that the reader should use extra care when analyzing the report.

Questions to consider when determining if a possible conflict of interest led to bias in a published study

- Did the introduction appear overly positive or unduly focused on the potential benefits of therapy?
- Were the inclusion/exclusion criteria slanted so that patients enrolled would be more likely to respond favorably to the study drug?

- Was an active control used that would lead clinicians to choose the study drug even if both are shown to be equally efficacious? Examples include if the active control is more expensive, not routinely used in practice, or known to have more adverse effects than the study drug.
- Was SEM provided for results when SD would have been more appropriate to include? Refer to the SEM and SD discussion in Chapter 9. Since SEM is always a smaller number than the SD, authors might include SEM with mean values simply to make the individual patient variability in their study appear less than it actually was.
- Did the authors discuss or interpret the findings in a way that was slanted towards or unduly emphasized positive (favorable) study findings? For example, did they downplay adverse effects or make nonsignificant differences appear more important than they actually were?
- Were statements made that were unsupported by the study's results, or did the authors inappropriately extrapolate their findings to situations not warranted by the study design/methods?

Study objectives and hypotheses

The rationale for performing a study should be clear from reading its introduction. The authors should have performed a thorough literature search to identify the previous work in that area and the knowledge "gaps" that their study will address. The literature cited should be balanced (to the extent possible) with studies that found favorable as well as unfavorable findings about the study drug. The introduction should also provide an accurate assessment of the potential benefits versus risks from the therapy based on the articles cited.

At the end of the introduction, studies should clearly and concisely state their objective (purpose) for performing the study or a more definitive hypothesis that provides a statement of the results expected. The objective or hypothesis is a critical starting point for a study because the design and methods chosen must be appropriate to fulfill the study's purpose.

Most studies usually report an objective instead of a hypothesis. Regardless, hypotheses are tested statistically to help fulfill the study objective. The *null hypothesis* assumes there is no difference among treatments or comparisons. Remember the null hypothesis because it is the basis for statistical testing (see Abate (2012) for more information about the null hypothesis in statistical testing). *Alternative hypotheses* indicate that a difference is expected between therapies. A *one-tailed* (*one-sided*) alternative hypothesis

Key Point

Most clinical trials studying a new therapy or new use for an established therapy are interested in seeing if changes occur in either direction, more efficacious or less efficacious. Thus, either two-tailed alternative hypotheses or objectives are given in clinical studies. An objective will state that the efficacy of a therapy is being determined without indicating that only improvement, or only worsening, is looked for in the findings. Most statistical tests used for data analysis in a clinical study will therefore be two-tailed as well.

states that the difference (change) will occur in only one direction, such as a studied drug will cause a greater effect, or a medication will cause less efficacy compared with another agent. A one-tailed hypothesis should be present in order to use one-tailed statistical tests appropriately for data analysis. Refer to Chapter 8 for more discussion about statistical tests. A *two-tailed (two-sided)* alternative hypothesis states that the difference (change) will occur in either direction, with one therapy being either more or less efficacious than another.

Worked example

Example 6.1

Investigators wish to compare the efficacy of a new drug to treat hypertension with enalapril, and state the objective of their study as: "The purpose of this study is to determine the efficacy of the new drug compared with enalapril for patients with mild to moderate hypertension."

Which of the following statements would best define the null hypothesis (statement of no difference)?

1. The new drug will be more efficacious than enalapril for the treatment of mild to moderate hypertension.
2. There will be no difference in efficacy between the new drug and enalapril for the treatment of mild to moderate hypertension.
3. The new drug will not be more efficacious than enalapril for the treatment of mild to moderate hypertension.

Answer

Statement 2 shows the null hypothesis. Statement 3 is not correct for this example since it implies that the new drug could in fact be less efficacious than enalapril. Statement 1 indicates a clear direction of effect and represents a one-sided/one-tailed alternative hypothesis.

Worked example

Example 6.2

The authors of a placebo-controlled study indicate their objective is to determine whether daily multivitamins have any effect on the development of common colds. The investigators received a grant to conduct this study from the manufacturer of One-A-Day multiple vitamins. The manufacturer also supplied the vitamins and placebos. Subjects are enrolled into the study and assigned to receive either daily One-A-Day vitamins or an identically appearing placebo for 12 months.

Any potential conflicts of interest?

Yes. The manufacturer of the vitamins being studied provided the study funding (grant), which creates a possible competing interest for the investigators. The grant does not automatically mean the study is not good, but rather the reader should carefully review this study when published to make sure there are no obvious biases present. It is not considered a competing interest for the manufacturer to supply the drugs/placebos used in a study if that is their only involvement.

Statement of null hypothesis?

The null hypothesis would be: "There is no difference between placebo and multivitamins on the development of common colds."

A summary of key points and how to apply the information from this chapter to practice follow.

Key Points

- Editorial boards and peer review are two methods to help ensure the quality of a journal and the studies they publish.
- There are several potential conflicts of interest (competing interests) that might affect the objectivity of a study's investigators.
- Financial conflicts are the most straightforward to identify and are usually focused upon.
- Journals should ask authors to disclose all potential conflicts of interest and clearly state these in a published study.
- The introduction of a study should clearly provide the rationale for why it is needed, based on a thorough analysis of relevant literature, and clearly state an objective or hypothesis.
- The null hypothesis is very important and is the basis for most statistical testing.

How to apply to practice

- Be very cautious when reading studies published in journals that do not have editorial boards or that do not use peer review; the chance for bias might be much greater compared to studies from journals that use these methods.
- Always look for potential conflicts of interest for the authors/investigators of a published study and, if one or more exist, carefully consider the various ways in which bias might be present.
- Determine if a study is adequately designed to fulfill its stated objective or hypothesis.

Self-assessment questions

Question 1

Investigators conducted a study comparing galantamine and placebo on quality of life in 60 patients aged 65–85 years with symptoms of mild Alzheimer's disease. The investigators state that this study is being conducted "to compare the efficacy of galantamine with placebo for quality of life in Alzheimer's disease patients." Two of the investigators are physicians employed by a university geriatrics center and one of the investigators is employed by Ortho-McNeil Neurologics (manufacturer of Razadyne brand of galantamine). Study funding came from a grant from Ortho-McNeil Neurologics. All three investigators own stock in Ortho-McNeil-Janssen Pharmaceuticals, and one investigator owns stock in Eli Lilly. Ortho-McNeil Neurologics provided identically appearing drug and placebo for the study. Are there any potential conflicts of interest for the investigators in this study? What would the null hypothesis be for this study?

Question 2
A study introduction describes how a shorter course of amoxicillin might be as effective as a longer duration of therapy. The investigators stated that the objective of their study was to compare the effectiveness of amoxicillin therapy for 3 days versus 8 days in adults with hospital-acquired pneumonia. What would the null hypothesis be for this study?

Question 3
You locate a study published in the *Annals of Pharmacotherapy* and find the following information on its website: "The Annals is an independent, peer-reviewed medical journal that advances pharmacotherapy throughout the world by publishing evidence-based articles on practice and research." Briefly describe what is meant by "peer-reviewed" in this example.

Question 4
The authors of a study stated the following: "We tested the hypothesis that melatonin could significantly improve the quality of sleep in patients with insomnia resulting from psychoactive drug therapy." Is the hypothesis in this study one-tailed or two-tailed?

Question 5
You are reading a published study comparing the efficacy of two drugs that states the following:

> From the School of Medicine, Duke University, NC. Funded by a grant from the National Heart, Lung, and Blood Institute, NIH. Study medications were provided by Pfizer (manufacturer of both drugs). Manuscript received February 16, 2009. Accepted June 30, 2009.

Are there any potential conflicts of interest for the study investigators based on the information provided? Does it appear that peer review was used for this article?

Question 6
You locate a study that has five authors listed: four of the authors are at the School of Medicine at the University of Washington and the fifth author is an employee of Abbott Labs, the manufacturer of the study drug. The study drug (drug A) is being compared to another drug (drug B) already marketed by another pharmaceutical company for treatment of the same condition as drug A. You observe that drug B is more expensive and not commonly used in clinical practice because of its cost and possible adverse effects. You also note that there are safe, less expensive, equally efficacious drugs available that could have been used for comparison with drug A in this study. What was a likely reason for the selection of drug B as the comparison drug in this study?

Summary

One of the initial steps in reviewing a study is to consider where it was published (the journal), who the authors/investigators were (any potential conflicts of interest), and why they conducted the study (the purpose). As discussed in this chapter, this information can provide valuable insight into potential limitations of the study before a more detailed critique is conducted.

7

Evaluating clinical studies – step 3: methods used

Learning objectives

Upon completion of this chapter, you should be able to:

- describe the importance of a study's eligibility criteria, the methods used for enrolling patients in a study (sampling), and informed consent
- discuss the advantages and limitations of each controlled experimental design and type of control
- explain the importance of random assignment in a study
- describe adherence and how it can influence a study's findings
- explain the importance of outcome measure selection in a study and the concepts of validity, reliability, sensitivity, and specificity related to the outcome measures used
- describe the dependent and independent variables in a study
- describe the scales (levels) of measurement of study data and the importance of differentiating among them.

Introduction

This chapter will focus on several important considerations when examining the methods used in a study, including the study sample and sample size; controlled experimental designs; assignment to treatment groups; blinding; drug treatments and adherence; and outcomes, variables, and measurements. It is nearly impossible to design the "perfect" study. Although study weaknesses and limitations might be present, however, one should differentiate those that could invalidate the findings versus those that simply limit the application of some results.

Eligibility (inclusion and exclusion) criteria

Eligibility criteria are used to define the characteristics of the subjects to be enrolled in a study. Those characteristics that must be present in order for the subjects to be included are referred to as *inclusion criteria*, and the

> **Key Point**
>
> Always refer to the patients/subjects enrolled in a study as the study *sample*. This makes it less confusing when learning how to interpret confidence intervals (refer to Chapter 9), since confidence intervals apply to the *population* outside the study sample.

> **Key Point**
>
> Eligibility criteria should also prevent outside interferences with the study and its methods to the extent possible, such as excluding:
>
> - nonstudy (concurrent) medications that might interact with the study drugs or have actions that could affect the condition studied
> - patients who have contraindications to the study drugs, such as allergies or impaired renal or hepatic function
> - patients who have other medical conditions that could interfere with the study findings or prevent them from successfully completing the study.

characteristics that would prevent individuals from being enrolled in the study are called *exclusion criteria*.

Why are inclusion and exclusion criteria important to examine in a published study? They tell us the *population* of interest that the study is targeting, meaning the type of individuals for whom the study's findings might be applied (extrapolated). Since it is virtually impossible to study an entire population of individuals, each trial ultimately enrolls a *sample* of the population that the investigators wish to examine.

The investigators want their study sample to be as similar as possible to persons in the desired population at large, so that their findings can be extrapolated to the population outside the study. One type of bias, *selection bias*, occurs when the study sample is chosen in a way that causes patients not to represent the desired population. For example, suppose patients of a certain ethnicity are more likely to have a specific medical condition. If a study of that condition selects its sample in a manner that underrepresents the at-risk ethnic group, selection bias could be present.

Worked example

Example 7.1

A study examined the efficacy of a new drug to treat hypertension. The inclusion criteria were: 40–75 years of age, normal renal function, and diastolic blood pressure between 90 and 105 mmHg. Patients were excluded if they had liver disease or were already receiving therapy for their hypertension. The new drug was found to be very efficacious in lowering diastolic blood pressure in these patients.

Can one assume that the new drug would also be efficacious in hypertensive patients with impaired renal function or liver disease?

No. The eligibility criteria specified that the patients must have normal renal and liver function. Since the efficacy of the new drug was not studied in patients with abnormal renal or liver function, one cannot extrapolate that the drug would work equally well in these patients; further study would be needed.

Sampling (enrollment) considerations

To obtain a study sample that best represents the enrollment of interest, everyone in the population, in theory, should have the same chance of being selected for the study. The best way to sample a population is through *random* methods in which chance alone determines who is selected. However, for experimental studies random sampling is not usually possible since it is difficult to access the entire population who has the medical condition or disease of interest.

Types of sampling include:

1. simple random – everyone in population is identified and a random procedure (e.g., computer-generated) is used to identify the persons for study inclusion
2. stratified random – when it is important that a study enroll similar numbers of patients who have or do not have certain characteristics (e.g., smokers versus nonsmokers, patients with diabetes versus without diabetes), the population is first divided into groups (strata) based upon the presence or absence of that characteristic, and then a random sample is taken from each group for study enrollment
3. cluster – all individuals present in identified, already existing "clusters" in the population are selected for study enrollment (e.g., everyone living in certain cities in a state, all persons attending certain hospitals in a region)
4. systematic – selecting every *n*th (e.g., 4th, 5th, 10th) person for study enrollment; if everyone in the population is known and the starting point is randomly selected, this is a type of random sampling
5. convenience – nonrandom sampling that enrolls patients based upon advertisements or whether they are being treated in certain clinics, hospitals or outpatient settings that the investigators work at or are affiliated with; used when it is not practical or possible to identify or contact all the members of the target population for study inclusion. *This is the method used most commonly in experimental studies* (e.g., a study of asthmatic children from a pediatric clinic who meet specific eligibility criteria, a study of drugs used to treat mild hypertension in patients from outpatient practice sites). This is an acceptable method for experimental studies as long as it is not known at the time of study enrollment which treatment group the patient will be assigned to (Bruce *et al.* 2008).

Worked example

Example 7.2

Investigators wish to study the efficacy of a new drug to increase smoking cessation, a condition in which subject motivation to quit is very important. Patients are enrolled who respond to a newspaper ad asking for volunteers who would like to participate in a study to quit smoking.

Could selection bias be a problem here?

Yes. The drug is anticipated to be more efficacious in motivated subjects. This study sample of volunteers responding to an advertisement is likely highly motivated to quit and might not represent the population of smokers at large. Thus, the results from such a study should only be extrapolated to a population of motivated smokers.

Informed consent

Investigators need to ensure that the subjects in their studies are protected from harm to the extent possible. To do this, an Institutional Review Board (IRB) is used along with the process of informed consent. IRBs were established as a result of federal regulations for the protection of human subjects that became effective in 1974. The history and responsibilities of IRBs can be found at the US Department of Health and Human Services, Office for Human Research Protections website (*Institutional Review Board Guidebook*; accessed at: http://www.hhs.gov/ohrp/archive/irb/irb_introduction.htm). An IRB is a group responsible for assuring that researchers in their organization/institution take appropriate steps to protect their subjects' rights and welfare, both before a study is initiated and throughout the study. An important part of protecting a subject's rights involves the need for investigators to obtain informed consent from each subject prior to enrollment in the study.

> *The informed consent process*
>
> According to the Food and Drug Administration, the following are part of the informed consent process (Food and Drug Administration, accessed at http://www.fda.gov/ScienceResearch/SpecialTopics/RunningClinicalTrials/default.htm):
>
> - providing a subject with adequate information about the study and its benefits and risks
> - giving the subject appropriate opportunity to consider all options
> - responding to the subject's questions
> - ensuring that the subject understands the information
> - obtaining the subject's written voluntary consent to participate in the study
> - providing additional information as needed.

With informed consent, the subjects in a study also have the right to quit the study whenever they wish. When reading a published study, keep in mind that with informed consent the patients in a study are aware of all the potential adverse effects and risks from each type of therapy they might receive.

Sample size

A common question when analyzing a study is, "Did the study have enough patients?" or "How many patients is 'enough'?" A study should have a sample size large enough to identify a statistically significant difference among treatments when an actual treatment effect exists. That is, it should be able to reject the null hypothesis, no difference among treatments, when it is indeed false and there really is a treatment effect. The extent to which a statistical test is able to identify a significant difference when there is an actual treatment effect is referred to as *power*.

Sample size is a key factor affecting a study's power. All else being equal, as sample size increases, statistical power increases. As the number of patients enrolled in a study decreases, the power decreases.

What's the problem with a low power?

If the statistical power is too low, a study could report a fairly large (and possibly clinically important) difference in outcome measures between treatment groups, but its analyses might not find this difference to be statistically significant. Thus, the concern is that a real treatment effect would be missed.

The study's sample size, or number of patients to enroll, should ideally be calculated before the study begins. To do this, investigators first select the desired power for the study. By convention, an acceptable degree of power is 80% or greater. Using the selected power value, they then calculate the appropriate sample size needed to achieve that power (using an equation for determining power and solving for sample size, one of the variables in that equation). The investigators should clearly state for readers the power of their study and the sample size needed to achieve that power. Refer to Chapter 9 for more discussion about power, its importance, and the factors affecting power.

Worked example

Example 7.3
A study reports that a total of 100 patients needed to be enrolled in each of two treatment groups to achieve a power of 80%. Although they are able to enroll a sample of 200 patients (100 in each group), several of these patients did not complete the study and only 80 patients were analyzed per group. The results showed there was a fairly large difference in the outcome measured between the treatment groups, but this difference was not found to be statistically significant. The investigators concluded there was no difference between treatments.

Were enough patients enrolled?
No. The sample size should be sufficient to achieve the desired power of at least 80%. Even though the investigators had enough patients at the start of their study to have 80% power, when the results were analyzed there were fewer patients. As the number of patients decreases, power decreases. With a power of less than 80%, there is a higher than acceptable chance that an actual treatment effect will be missed. Thus, insufficient power might have been the reason why this study missed a fairly large difference between treatments.

Controlled experimental designs

The controlled experiment is a strong design for proving cause and effect and establishing therapy efficacy. *Controls* (comparison groups) should be used when possible since they reduce the likelihood that outside factors that are not part of the study (e.g., environment, nonstudy medications taken by the groups) might affect the study results. Chapter 5 briefly defined the types of control groups: *placebo, active, historical, no treatment*. Three primary designs (Figure 7.1) are used for controlled experiments: (1) concurrent control (parallel); (2) cross-over; and (3) time series (before and after).

In the *concurrent control* or *parallel* design, patients are assigned to receive either a control or study treatment. Patients only receive one of the interventions with a parallel design. Results are then compared between/among groups to determine if the outcomes in one group are significantly different than in the other. With the parallel design it is important that the patients in each group are as similar as possible to help ensure comparability of the results.

In a *cross-over* design, the patients receive each of the interventions (control and treatment). They are initially assigned to either the control or experimental therapy. After completing that course of treatment, the patients are assigned to the other group(s), one at a time, so that each patient eventually receives each intervention by the study's end. The cross-over design generally includes a wash-out period, an amount of time during which no therapy is given, between each intervention so that (in theory) any effects from one treatment can be eliminated or "washed out" from the body prior to beginning the next intervention phase.

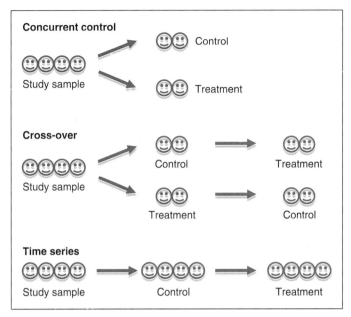

Figure 7.1 Illustration of controlled experimental designs.

Key Points

- The type of control selected, whether placebo or active, should be appropriate for the purpose of the study.
- The concurrent control design is preferred since its advantages generally outweigh the disadvantages.
- The time series design is least desirable because, unlike a cross-over design, effects of time and drug sequence on the study findings cannot be determined.

In the *time series* or *before-and-after* design, each patient also receives each study intervention. However, this design differs from the cross-over design in that all patients receive the same type of intervention at the same time. The advantages and disadvantages of each of the three controlled experimental designs are summarized in Table 7.1.

With a cross-over design, treatment effects can be compared between and within periods to determine if the response to an intervention during one period is the same as the response to that same intervention during the next period. This involves comparing: (1) the difference in response between the treatment and control in the first period with the difference in response between both in the second period; and (2) the control effect during the first period with the control effect during the second period, and the treatment effect during the first period with the treatment effect during the second period. If there is no carry-over or other sequencing effects, those comparisons should all be the same. If any differences are present, the results can be difficult to interpret because it is

Type of controlled design	Advantages and disadvantages
Concurrent control/parallel	Advantages: • Most straightforward to analyze statistically • Effects from one study intervention cannot carry over and affect results from other intervention(s) • Requires less time than cross-over or time series designs • Time itself affecting condition studied is less likely compared with other designs Disadvantage: • Are patients in different intervention groups comparable? Clear eligibility criteria and random assignment are important to help assure patients in different groups are similar
Cross-over	Advantages: • Easier than parallel design to eliminate patient differences in the intervention groups as the cause of any effects seen since the same patients receive each intervention • Requires fewer patients than concurrent control design for a desired power • Unlike time series design, can analyze findings to help determine if carry-over or other sequencing effects occurred Disadvantages: • Requires more time to complete than concurrent control design • Wash-out period important to help reduce possibility of treatment carry-over but cannot guarantee it will prevent carry-over • More complex statistical analyses needed to determine possibility of carry-over or other period/sequencing effects
Time series/before-and-after	Advantages: • Easier than parallel design to eliminate patient differences in the intervention groups as the cause of any effects seen since the same patients receive each intervention • Requires fewer patients than concurrent control design for a desired power Disadvantages: • Requires more time to complete than concurrent control design • Wash-out period important to help reduce possibility of treatment carry-over but cannot guarantee it will prevent carry-over • Cannot analyze results to determine if a carry-over or other sequencing effect might have occurred

Table 7.1 Advantages and disadvantages of controlled experimental designs

unknown how much of the effect was due to the treatment versus carry-over or another nondrug influence.

Worked example

Example 7.4

Investigators conducted a randomized, single-blind, placebo-controlled study of galantamine (G) and rivastigmine (R) given for 12 weeks in 60 patients aged 65–85 years with mild Alzheimer's disease. The investigators state that this study is being conducted "to determine the efficacy of G and R on quality of life in Alzheimer's disease patients."

How would this study be conducted using a parallel study design? How would it be conducted using a cross-over design?

With a parallel design, the patients would be randomly assigned to receive either G or R for 12 weeks. They would only receive this one therapy. With a cross-over design, the patients would be randomly assigned to receive either G or R first for 12 weeks. A wash-out time period of no drug administration should then occur, followed by the patients receiving the drug they did not already receive, G or R, for 12 weeks. Since patients receive each of the therapies in a cross-over design, notice that the total study duration is longer.

Assignment to interventions

Key Points

- Any time a study states it is "randomized," this is referring to random *assignment* to treatment groups.
- With random assignment, patient characteristics (known or unknown) that might affect the study results are more likely to be similarly distributed between the study groups.
- Random assignment will *not* guarantee that patients in the different study groups will have identical characteristics. By chance, there still might be important study group differences that could influence the results.
- To determine the success of randomization, investigators will usually include in a table the demographics (e.g., age, gender, race) and key baseline (prior to therapy) characteristics of the patients assigned to each study group. These findings should be compared and analyzed to determine if the groups are indeed similar with regard to factors that might alter therapy response.
- If important baseline differences exist, statistical analyses might be able to take these into account when interpreting findings.

How should patients/subjects be assigned to a certain intervention group, e.g., control versus treatment? If an investigator simply chooses who receives each intervention, bias could be present (consciously or subconsciously). Sicker patients might be assigned to one group over another. Thus, patient assignment is best done through a process called *randomization*, in which patients are randomly assigned, e.g., using computer-generated random numbers or a random numbers table, to their study group. Random assignment means that each patient has an equal chance of being in a particular study group. This is important to help reduce bias and ensure study quality. Random assignment also reduces the chance that any extraneous factors might affect the study results since differences in patient characteristics would likely be "balanced" between groups.

Blinding

Blinding, or *masking*, is when the patients/subjects and/or investigators do not know the intervention group that the patients were assigned to. Thus, it is not known during the study whether a patient is receiving the control or the study drug. Blinding should be used to reduce the risk of bias that might result from patients or investigators knowing who is receiving which therapy. Many outcome measures, particularly those that are subjective in nature, such as pain relief, changes in mood, or development of side-effects, can be affected by a person's belief (patient or investigator) that a therapy will work or have a certain effect. An investigator might be hopeful that a new drug will be efficacious and unintentionally skew the results in that direction by asking patients more questions about the drug or repeating tests if findings are unexpected. If patients believe that the drug they are taking will be beneficial, they might report changes they would otherwise overlook or fail to mention potential problems experienced.

> *Definitions*
>
> Generally used definitions for the types of blinding include:
>
> - *Single-blind* – the patients (usually) are unaware of the therapy they are receiving but the investigators know
> - *Double-blind* – neither patients nor investigators (assuming they perform the analyses) know which therapy each patient is receiving
> - *Triple-blind* – if any noninvestigators perform the analyses, neither they, the patients, nor investigators know which therapy each patient is receiving
>
> In an open-label or nonblinded study, both the patients and investigators know which therapy each patient is receiving.

> *Can different dosing frequencies, e.g., once daily versus twice daily, be blinded? Is it possible to blind a study that is comparing two different dosage forms, e.g., tablet versus capsule?*
>
> Both can be done. The once-daily and twice-daily drugs should look alike. Patients taking the once-daily drug would receive an identical placebo for the second dose.
>
> The *double-dummy* method is used to blind different dosage forms. If comparing a drug formulated as a tablet with a drug in a capsule dosage form, the patients in the active tablet group would take their tablet with a placebo capsule, and the patients in the active capsule group would take their capsule with a placebo tablet. The same "double-dummy" approach could be used for any combination of dosage forms.

In a single-blind trial, the investigators might slant the study's measures in a certain direction or lead patients to figure out the treatment they are receiving. Thus, double-blind trials are preferred to minimize the likelihood of bias and are an important part of the "gold standard" randomized, controlled experimental study design. However, it is not always possible to double-blind certain interventions in a study. For example, one could not blind a comparison of the effectiveness of indepth patient counseling to minimal patient counseling or a drug to surgical therapy. A study might also start as double-blind, but certain side-effects (e.g., nitroglycerin headache), unique drug or treatment smells/tastes (e.g., a study of garlic's blood pressure effects), or characteristic lab test alterations "clue in" the patients or investigators to what the patient is actually receiving. *Unblinding* (unmasking) occurs when the patients or investigators can successfully identify what the patient is receiving during a blinded study. If unblinding occurs to a significant extent, the potential benefits of a blinded study disappear.

How can one tell if unblinding occurred?

In a study comparing two different treatments, at a minimum the treatments should look alike for the study to be double-blind. When considering the possibility of unblinding, think of other possible reasons why it might occur, such as a drug's distinctive taste, odor, and side-effects. Would nonhealth professional patients know the possible side/adverse effects of their study drug? Yes. Remember that, through informed consent, patients are told the potential benefits and risks (including side/adverse effects and their likelihood) for each study treatment they might receive.

The simplest way to determine if unblinding occurred is to ask patients which therapy they received. If most patients can successfully guess their treatment, unblinding was likely. Keep in mind, however, that if a drug is highly efficacious compared to a control, such as placebo, patients and investigators might correctly guess their therapy simply by the patient having a good response. Thus, determining unblinding by asking patients is somewhat questionable at times even though it is generally the only available option.

Worked example

Example 7.5

A double-blind study compared a new nonsteroidal anti-inflammatory drug (NSAID) to placebo for the management of arthritis pain. Patients who provided informed consent were randomly assigned to receive either the NSAID (80 patients) or placebo (65 patients) for 10 weeks. Identically appearing NSAID and placebo tablets were used. The NSAID was found to be significantly more efficacious than placebo in relieving pain. Adverse effects were reported by 75% of patients receiving the NSAID (nausea, stomach pain, and headaches were most frequently reported) compared to 15% of placebo patients.

Would unblinding be of concern in this study?

Yes. Although the drug and placebo looked alike, there were many more adverse effects with the NSAID that could have led patients to guess correctly the treatment they were receiving. With informed consent, patients are told the possible adverse effects of each study treatment they might receive, so patients would be able to recognize the NSAID adverse effects. The investigators could have checked for possible unblinding during and at the end of the study by asking the patients to guess the study treatment they were taking. If the majority of patients could correctly guess their treatment, unblinding was likely.

Treatment considerations

There are several aspects to consider when reviewing the drug(s) being used in a clinical study, including:

- dosage and dosage forms
- dosing frequency
- route of administration
- duration of therapy
- drug concentrations obtained
- use of any concurrent (concomitant) nonstudy medications
- adverse effects
- therapy adherence (compliance).

The dosages of the experimental drug and any active control in a study should be appropriate for the drugs used. If there are established dosing ranges, the drugs should best be dosed at comparable ends of the ranges, e.g., both dosed at the lower end or both dosed at the higher end. In general, established drugs should be dosed similarly to how they would be dosed in practice: in a single, fixed dose or individually adjusted doses, as most appropriate.

The dosing frequency should be consistent with the known pharmacokinetics of the study drugs. If a drug has an established therapeutic concentration range, the study should ensure that concentrations are measured in patients and are within that range. Concentrations should be measured at appropriate times of the day and at steady state for chronic therapy. The duration of therapy should be long enough to allow for the full therapeutic effect to occur and for adequate time to assess the benefit on the medical condition.

> **Key Point**
>
> If a concurrent medication might affect the study outcomes, determine whether:
>
> - comparable quantities were taken among the patients in each study group
> - the overall amounts taken were large enough to impact the findings.
>
> For example, suppose a study comparing a new antihistamine to placebo does not exclude decongestant use. Since decongestants could affect certain allergy symptoms, their use could alter some of the study findings. In this study, the investigators should record the amount of decongestants taken in both the antihistamine and placebo groups. The reader should ask:
>
> - Were decongestants used to a large enough extent in both groups to cause significant reductions in nasal congestion symptoms?
> - If decongestants were taken by more patients in one of the groups, could they have made the antihistamine appear more (or less) efficacious?

A study may or may not allow patients to take concurrent, nonstudy medications. When outside medications are known to interact with the study drugs or will otherwise affect the condition being studied, they will usually be excluded as part of the exclusion criteria. If concurrent, nonstudy medications are allowed during a trial, consider whether they could interact with the study drugs or affect the disease state or symptoms studied.

Adverse effects should always be considered when determining the benefit versus risk of a therapy. Mild as well as serious adverse effects can be of concern and affect the clinical usefulness of a drug. For example, a drug can be very efficacious but cause a large number of minor, but annoying, side or adverse effects. Another drug might be very efficacious and cause only a few adverse effects, but the effects are serious. Both situations could result in patients dropping out of a study or being nonadherent (noncompliant) with their therapy. As previously mentioned, adverse effects are also a source of unblinding in a study.

Patient adherence can be an issue for both the procedures used in a study (e.g., completing a diary of symptoms or adverse effects, filling out diet/meal cards) as well as the drug therapy. A study will not be able to obtain complete, accurate data if patients do not complete needed diaries or records. Likewise, adherence with a therapeutic regimen is very important in clinical studies and in practice. A therapy cannot be expected to work if the patient does not take it. It is important that a study always considers adherence for any needed patient-supplied diaries/records and the drug therapy, and assesses the extent to which it occurred. Otherwise, the results seen might reflect varying adherence rates and not actual differences in efficacy. *Compliance bias* occurs when a difference is seen between study groups as a result of differences in adherence rates rather than an actual difference in drug effects.

Key Point

A study should measure and compare the adherence rates to the therapy (and other necessary study procedures) in each patient group to determine if compliance bias might have occurred. Adherence rates should be similar to rule out possible compliance bias.

It is easy to measure adherence when patients are closely monitored (e.g., inpatient studies, intravenous medications administered by nurses) or when patient participation in the intervention can be easily determined (e.g., attendance at counseling sessions or scheduled office/clinic visits, undergoing surgical procedures). For adherence with study procedures, diaries or other records can be reviewed and missing entries quantitated. Measuring adherence to drug therapy can be more difficult and less reliable, particularly in outpatient settings in which the patients are responsible for taking all their drug doses. Since there is no perfect method for assessing drug therapy adherence, it is best for a study to use more than one approach.

Methods for determining adherence

For more information refer to *Fundamentals of Clinical Trials* (Friedman et al. 2010).

- Pill counts – quick and easy but can be unreliable (inflated) since doses can be lost or nontaken doses thrown out.
- Review of pharmacy refill records – easy but obtaining a prescription does not indicate whether patients actually took the drug as they were supposed to.
- Use of electronic caps or devices that record the times the drug's container or package were opened – can be a good method when available.
- Asking patients directly or have patients keep a diary/record of doses taken – can be inaccurate.
- Measuring drug concentrations or adding an inert "marker" to a drug product that can be measured – can indicate if a drug was taken, but not adherence with all doses or with the dosing schedule since missed doses could be taken shortly before a scheduled visit.
- Measuring a drug's physiologic effect on a nonprimary study outcome (e.g., certain lab changes, antibacterial activity present in urine) – interpreting this information can be difficult since patients might respond differently to these variables and it does not guarantee that patients took the drug regimen the way it was prescribed.

Worked example

Example 7.6

Two pain medications, drug A and drug Y, were compared for the treatment of patients with severe back pain. After 3 months of therapy, 89% of patients receiving drug A reported "complete" pain relief compared to only 60% of drug Y patients. Bitter taste was reported by 80% of patients receiving drug Y; other side-effects included mild nausea and headache in 4–5% of patients taking either drug. Through the use of pill counts by investigators and diaries that the patients used to record their doses taken, 96% of patients receiving drug A took at least 90% of their doses compared to 65% of drug Y patients who took at least 90% of their doses.

Was adherence appropriately measured in this study? Was compliance bias a potential problem?

There are several methods that can be used to determine the degree of patient adherence in a study; none of them is 100% accurate so it is best to use a combination of methods. This study used two methods (pill counts, patient diaries), which is acceptable. Compliance bias occurs when different rates of patient adherence in the treatment groups lead to differing efficacy rates. The much lower degree of drug adherence by the drug Y patients could explain the finding of less efficacy of drug Y compared to drug A.

Outcomes and variables

A study's objective should specify the broad, overall outcome(s) of interest, e.g., hypertension control, diabetes control, smoking cessation. In their methods, the investigators should clearly specify their *primary outcome(s)* of interest as well as any *secondary outcomes*, i.e., still of interest but not the main concern or focus of the study. Studies should also indicate a desired end point where appropriate, meaning the point at which an outcome will be successfully met or the study hypothesis supported. For example, in a study comparing the efficacy of two antihypertensive drugs, the primary outcome might be diastolic blood pressure changes and an appropriate end point might be a diastolic blood pressure of <80 mmHg. Secondary outcomes might include systolic blood pressure, renal function, serum cholesterol, or other metabolic changes that could result from the therapy.

Variables are related to the outcomes and refer to a study characteristic that can assume different values. Two important study variables are the *dependent* and *independent variables*. The independent (explanatory) variable(s) in a study affect the value of the dependent (response) variables. In a clinical study, type of treatment is generally *one* independent variable. The dependent variables in a study are those that change in value as a result of the independent variable; in a clinical study, the outcome measures are the dependent variables since their values would be altered by exposure to the therapy (control or treatment) administered.

A third type of variable, the *confounding variable*, can also play an important role in studies. It is a factor that can affect the value of the outcome measures in addition to the therapy being studied. For example, suppose a study compares a new antiviral drug to placebo to prevent influenza. Several study patients had recently been vaccinated with the influenza vaccine. The vaccine would be a confounding variable that could affect the determination of drug efficacy. Investigators need to take such variables into account because they can interfere with the study results and their interpretation.

Measurements

The tests or procedures used to measure changes in the desired outcomes (dependent variables) should be appropriate for the intended purpose. Ideally, they should measure only the effect of interest without allowing interference from unrelated factors. Studies should select the best test or combination of tests to measure an outcome, such as using a "gold standard" test when appropriate and feasible for the situation and study setting.

For a study's measurements to be meaningful and truly indicate the extent to which the outcomes have been achieved, the tests or procedures used should be: (1) valid; (2) reliable; (3) sensitive; and (4) specific.

> *Important characteristics for measurements*
>
> - *Validity* – the extent to which a measure is truly determining what is desired or what should be measured.
> - *Reliability* (reproducibility, consistency) – the extent to which a measure of the same outcome provides similar results when used at different times or by different individuals.
> - *Sensitivity* – the ability to which a measure can identify or detect the presence of a condition or characteristic when present.
> - *Specificity* – the degree to which a measure can accurately identify as negative those who do not have a condition or characteristic; or the extent to which a measure can accurately detect only the condition or characteristic of interest.

Worked example

Example 7.7

Suppose the Arthritis Quality of Life Scale is able to measure accurately only the changes in quality of life that result from arthritis but not from any other medical conditions. It is not able to detect small changes in quality of life, however; it only detects fairly large changes.

> **Based on the information provided here, which of the following characteristics, validity, reliability, sensitivity, or specificity, does this survey instrument appear to possess?**
> Specificity, since it measures quality of life changes only due to arthritis, and not other conditions. The Arthritis Scale also seems to be valid if it accurately measures the quality of life. However, one cannot determine the reliability of this instrument given the information provided. For scales, surveys or questionnaires that are developed by investigators, it is very important that their validity, which includes reliability, be determined (i.e., the scale, survey, or questionnaire should be *validated*) prior to use in a study. Sensitivity seems to be lacking in that the scale misses small changes in quality of life.

There are several types of validity, including internal, external, face, content, construct, and criterion validity. Two of the major ones to consider when evaluating a study are internal and external validity. *Internal validity* is the extent to which a study's findings were appropriate and correct – was the relationship found between the intervention and outcomes accurate? If a study has internal validity, the extent to which its findings can be applied or generalized to patients and settings outside the study is referred to as *external validity*. For example, if a study enrolled patients who did adequately represent the population of interest or used a treatment dosage not normally used in practice, the external validity will be limited. Additional information about the types of validity, reliability, sensitivity, and specificity can be found in other references such as *Quantitative Methods for Health Research* (Bruce et al. 2008).

Key Points

- The stronger a study's design, methods, and analyses, the greater the internal validity. Randomized, controlled experimental studies have stronger internal validity than other types of trials. However, any important weaknesses or flaws will reduce internal validity, even with a "gold standard" study design.
- External validity (generalizability) is an important consideration for clinicians interested in applying the results from a study to their patients.
- A variety of factors can affect external validity, including the types of patients enrolled and how they were selected, the study setting (e.g., inpatient versus outpatient), and even the timing of the study (e.g., studying antihistamines at a time of the year when certain allergens might be less prevalent).
- A measure can be reliable without being valid, but to be valid it must also be reliable.
- A study should provide any needed training and use clear, standardized directions, protocols, and definitions to ensure that an outcome measure is appropriately used across study locations and by different investigators and patients.
- Patients should also be handled similarly, with the exception of the treatments used, to help ensure that the study methods or measures themselves did not lead to altered patient behaviors that might impact the study. The *Hawthorne effect* is when patients alter their performance, behaviors, or attitudes simply as a result of being observed or given attention in a study, *not* from the intervention given.
  - *How can the Hawthorne effect be minimized in a study?*
    It can be reduced by handling patients in each treatment group as similarly as possible, other than the treatment given.

> For example, if one study intervention required patients to be seen by investigators each week in the clinic for follow-up visits while the other intervention only required patients to be seen once monthly, the patients being seen more frequently might alter their behaviors just because they are being observed more often. Suppose these patients were in a study comparing two antihypertensive drugs. Patients being observed more closely might decide to reduce their salt intake and exercise to a greater extent, simply as a result of being observed more closely. This behavior change could make their drug appear more efficacious, when reduced sodium and greater exercise alone were responsible for the difference.

Measurements produce numerical data that can have different meanings. For example, the number 5 could indicate the number of patients with a certain characteristic, a value in a rating scale that means "extremely positive," or a serum potassium concentration. *Levels* or *scales of measurement* differentiate these meanings. The level/scale of measurement of specific data is very important because it determines the type of statistical test to use for the data analysis. The three levels of measurement are nominal, ordinal, and continuous, with continuous data further divided into interval and ratio levels. Since the same types of statistical tests are generally used for interval and ratio-level data found in clinical studies, the term "continuous data" will be used hereafter to include both.

> *Definitions of levels/scales of measurement*
>
> - *Nominal (categorical)* – data that lack numerical qualities but can be placed in mutually exclusive categories; examples include:
>   - gender
>   - race
>   - presence or absence of adverse effects or outcomes, such as the proportion/percentage of patients with a headache, those who were cured, or those whose hypertension was controlled
> - *Ordinal* – data that can be rank-ordered on a scale; one value is more or less than another, but any assigned numbers do not have exact differences between them (e.g., the difference between a 1 and a 2 is not exactly the same as the difference between a 2 and a 3); examples include:
>   - opinions ranked using a scale of 5 = strongly agree, 4 = agree, 3 = neutral, 2 = disagree, 1 = strongly disagree
>   - ranking of quality of life as 4 = excellent, 3 = good, 2 = fair, 1 = poor
>   - rating of the severity of a condition as severe, moderate, mild, absent

- *Continuous* – data that can assume an unlimited number of numerical values within a range, with equal distances between numbers; the types include:
  - *interval* – continuous data that lack a true zero point; cannot be presented as a ratio; examples include Fahrenheit or Celsius temperatures and pH values
  - *ratio* – continuous data that have a true zero point; it is the most common type of continuous data used for clinical research; examples include height, weight, and drug concentrations.

Worked example

Example 7.8
List the level/scale of measurement for each of the following types of study data:

1. Thyroxine serum concentrations following thyroid hormone replacement
2. The severity of neuropathic pain (3 = severe, 2 = moderate, 1 = mild, 0 = absent) following therapy with gabapentin or placebo
3. Number of osteoporosis patients who experienced a fracture during treatment with either alendronate (3 of 29; 10.3%) or risidronate (5 of 35; 14.3%)
4. Hemoglobin A1c concentrations at baseline and following metformin therapy.

Answers

1. Continuous – there are an infinite number of possible values, with equal distances between numbers.
2. Ordinal – the severity can be rank-ordered, but the difference between a pain score of 1 and 2 is not exactly the same as the difference between 2 and 3.
3. Nominal – the patients can be placed into mutually exclusive categories: experienced a fracture or did not experience a fracture.
4. Continuous – there are an infinite number of possible concentration values, with equal distances between numbers.

A summary of key points and how to apply the information from this chapter to practice follow.

Key Points

- The inclusion and exclusion criteria should accurately characterize the type of patients that a study wishes to target.
- Apply the findings from a study's sample to the population specified by the inclusion and exclusion criteria.
- Random sampling (selection) is the best method for enrolling subjects in a study, but this is not usually possible for experimental studies.

- The sample size should be large enough to have sufficient power to identify a difference in outcome measures as statistically significant if there is a real treatment effect.
- A study's power should be at least 80%.
- Investigators are required to obtain IRB approval before initiating a study, which includes obtaining informed consent prior to participants' enrollment in the study.
- The concurrent control (parallel) experimental design is best and is preferred; the cross-over design is better than the time series design.
- Double-blinding (or triple-blinding, if needed) and randomization should be used whenever possible.
- If an active control is used it should be an appropriate choice to compare to the therapy being studied.
- All drugs being studied should be dosed and administered appropriately.
- If concurrent nonstudy medications are allowed, the use of any that might affect the study treatments or results should be quantitated and compared between groups.
- Adherence should always be considered in a study and reported. More than one method for determining therapy adherence should best be used.
- Primary and secondary outcomes should be clearly defined and appropriate for the study objectives.
- Study measurements should be valid, reliable, specific, and sensitive.
- The level/scale of measurement of data is an important factor that determines the type of statistical test to use.

How to apply to practice

When evaluating a study, be sure to check:

- the inclusion and exclusion criteria to determine the population to whom a study's results can be applied, and whether your patients (and the population at large) are comparable to those sampled in the study
- the number of study patients needed to reach a power of 80% or greater. Make sure that the number of patients *analyzed* did not drop below that needed for adequate power. If the power was less than 80% or never stated, it is possible that a real treatment effect was missed for nonstatistically significant findings, especially if the difference between groups appeared fairly large
- the dosing and administration regimens used to ensure they are appropriate and representative of what would be used in clinical practice
- whether sufficiently long wash-out periods were used between the interventions in a cross-over study; the investigators should also have analyzed the results for carry-over or other period or sequencing effects
- the baseline (postrandomization) characteristics of the patients assigned to each study group to see if:
  - potentially important factors are evenly distributed among groups. If not, consider whether these differences might have altered therapy response in the groups
  - there were important baseline comparisons that the investigators overlooked. Examples might include differences among groups in duration or severity of illness, previous therapies tried and those currently being taken, and underlying renal or hepatic function

- the possibility of unblinding in a blinded study
- whether concurrent nonstudy medications could have significantly affected the findings
- whether adverse effects led to significant patient drop-out or nonadherence, were frequent or serious enough to affect the clinical usefulness of the drug, or might have led to unblinding
- the adherence rates for each study group. If significant adherence differences are present among study groups, ask whether these differences might have contributed to the results seen (e.g., did the group with much lower therapy adherence also have a lower drug response?)
- whether the validity and reliability were determined for any surveys or questionnaires that the investigators developed themselves
- if the outcome measures used were appropriate and optimal given the study objectives
- whether any limitations or weaknesses in a study or its design could reduce the internal validity, thereby affecting the external validity.

Other considerations:

- If an *observational* study cannot include the entire population of interest, random sampling methods should be used if possible to minimize bias in subject selection.
- Don't be concerned if an experimental study uses nonrandom convenience sampling. This is acceptable as long as the patients are randomly assigned to a treatment group following enrollment.
- Watch for differences in how the patients were handled that might affect the study results independently of the treatments used (i.e., Hawthorne effect).

Self-assessment questions

Question 1
Name the *type of validity* that refers to the ability to extrapolate a study's results to other patients who were not part of the study.

Question 2
A randomized single-blind study compares a new triptan (T) drug to sumatriptan (S) to treat migraine headaches. A total of 54 patients are assigned to receive either T or S. When a migraine is experienced, patients take a dose of their assigned drug and record the severity of pain over the next 24 hours. The outcome measures include the time to headache relief and the average pain severity over the 24-hour period. Patients are allowed to take prn NSAIDS if they do not experience headache relief within 2 hours after their T or S dose. What do the terms "single-blind" and "randomized" indicate in this study? Would the use of NSAIDS be a potential problem in this study?

Question 3
An open-label study was performed to compare the effect of cyclobenzaprine with other muscle relaxant drugs on tinnitus severity in patients who have chronic tinnitus. Cyclobenzaprine, but not the other treatments, was found to improve tinnitus severity

significantly. What does "open-label" refer to in this study, and can it have an effect on the study findings?

Question 4

A randomized, double-blind, controlled study was conducted to evaluate the efficacy of Viagra (sildenafil) compared to Lyrica (pregabalin) for treatment of neuropathic pain. A total of 150 patients who presented to one of three pain clinics affiliated with the University of Southern California from January 2003 to December 2004 and who met eligibility requirements were enrolled in the study. Would it be correct to state that this study used consecutive random sampling to enroll its patients?

Question 5

What measurement term refers to the ability of a study to accurately find true and correct results *within* the study based on having a strong design and methods used?

Question 6

A study of methotrexate for treating rheumatoid arthritis measured the degree of joint erosion and joint space narrowing on X-rays as two of the primary outcomes. Assume that the X-rays can measure even very small changes in the degree of joint erosion and joint space narrowing. Which measurement attribute describes this ability to detect very small changes in the outcome measure?

Question 7

A study is performed to compare metformin to glipizide for diabetes therapy in type 2 diabetic patients. A total of 120 patients are randomly assigned to receive metformin or glipizide for 12 weeks. At the end of the 12 weeks, the patients are then switched to the other therapy for an additional 12 weeks. Which controlled experimental design was used in this study? Is this the preferred design?

Question 8

A double-blind, randomized study examined the use of a new medication to treat patients suffering from psoriasis. A total of 28 study sites were used at various locations in the USA and Canada. One of the outcome measures included the rating of psoriasis severity by investigators using the following rating scale: 4 = severe; 3 = moderately severe; 2 = moderate; 1 = mild; 0 = absent. The study found that, following use of the medication, there was a slight but significant reduction in the rating of psoriasis severity. Are there any potential problems with the rating scale data from this study?

Summary

The methods section is one of the most important parts to read in a published study. A study must be appropriately designed in order to fulfill its hypothesis and objectives, and the methods section specifies precisely how the study was conducted. Serious weaknesses in a study can invalidate its findings. Several important study design considerations were covered in this chapter. Readers need to be familiar with these considerations, including the use of controls, randomization, and blinding, among others, in order to analyze and draw valid conclusions from clinical studies.

8

Evaluating clinical studies – step 4: statistical analyses

Learning objectives

Upon completion of this chapter, you should be able to:

- indicate when one-tailed versus two-tailed statistical tests should be used
- define parametric tests and describe the main features of the following commonly used parametric tests: t-tests and analysis of variance (ANOVA)
- define nonparametric tests, when they should be used, and the main features of the following commonly used nonparametric tests: Chi-square, Fisher's exact, Mann–Whitney U, Wilcoxon signed-rank, Kruskal–Wallis, Friedman
- describe the purpose of correlation and regression analyses, including commonly used correlation tests and their interpretation.

Introduction

The interpretation of a clinical trial's findings usually depends upon the statistical analyses of the data generated in the trial. Statistics help the investigators, and readers, answer three types of questions:

1. Are any group differences found in the outcome measures likely to have resulted from the treatments, or from something else such as chance?
2. Are there associations among variables measured in the study groups?
3. What predictions can be made for the population based upon the results obtained from the study sample?

Many statistical tests are available; their selection depends upon the types of data and specific conditions involved. This chapter provides an overview of several of the more commonly used statistical tests. The discussion will focus on understanding the appropriate use of these statistical tests, rather than on involved mathematical calculations. For further information about statistical analyses used in clinical trials, the reader is referred to *How to Report Statistics in Medicine* (Lang and Secic 2006), *Intuitive Biostatistics* (Motulsky 2010), and Bolton and Hirsch (2012).

One-tailed versus two-tailed tests

> **Key Point**
>
> A clinical study should only use a one-tailed statistical test if it clearly stated a one-tailed hypothesis. A one-tailed hypothesis should only be used if the investigators are very confident of the result or have other evidence of a unidirectional change.

In Chapter 6, null and alternative hypotheses were discussed, with alternative hypotheses being either one-tailed (an expected direction of the effect is stated) or two-tailed (a change is expected but it can be in either direction, improvement or worsening). Most clinical studies involve testing new therapies or new uses for established treatments; thus, it is possible that improvement or worsening could occur. As a result, clinical trials generally tend to use objectives or two-tailed hypotheses that do not specify a specific change. The use of one-tailed (one-sided) or two-tailed (two-sided) statistical tests depends upon the study's stated objective and/or hypothesis. Since most clinical studies look for therapy changes in either direction, they will use two-tailed statistical tests.

Categories of statistical tests

There are two broad categories or types of statistical tests, parametric and nonparametric; each of these categories includes a number of individual tests. The choice between a parametric or nonparametric test to analyze study data depends upon the assumptions made about the underlying population from which the sample data were selected. *Parametric tests* are used when the data being analyzed are continuous, with the assumption that the underlying population is *normally* (or *near normally*) *distributed*. A normal (also called *Gaussian*) distribution resembles a bell-shaped curve when graphed by frequency. With a normal distribution, most individual data points are found near the middle, with roughly equivalent numbers of data points lying on each side of the midpoint, with decreasing numbers seen as one moves further from the midpoint (Figure 8.1).

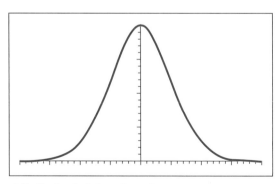

Figure 8.1 Normal distribution (bell-shaped curve).

> **Key Point**
>
> Three primary factors should be considered when selecting a statistical test or determining whether one was used appropriately in a study:
>
> 1. the level/scale of measurement of the data being analyzed (refer to Chapter 7 for a discussion of levels/scales of measurement)
> 2. how many treatment groups are being compared
> 3. whether the data are from paired or nonpaired (unpaired) patients.
>
> *Nonpaired* (*unpaired*) refers to data from different patients. *Paired* refers to data from the same patient, such as:
>
> - after receiving each treatment in a cross-over or time series study
> - before treatment (at baseline) and after treatment in the same patients in a parallel, cross-over or time series study
> - when patients are matched pairs, i.e., enrolling a patient in one treatment group with a "matched" patient in another group who shares similar characteristics.

A *nonparametric test* should be used when the study data are nominal- or ordinal-level, with no underlying assumption of normal distribution. Nonparametric tests are also used when at least one of the assumptions for use of parametric tests is violated/not true; most commonly this is when continuous-level data are not normally distributed.

Parametric tests are preferred when appropriate because they have more statistical power than nonparametric tests (i.e., ability to identify a treatment difference as statistically significant if there really is a treatment effect), particularly with relatively small sample sizes.

Parametric statistical tests

Two of the most commonly used parametric tests in the medical literature are the (Student's) *t-test* and *ANOVA*. In addition to the need for continuous, normally (or near normally) distributed data for use of a t-test or ANOVA, two other assumptions exist: population variances are equal (or nearly equal), and the observations or measurements within a population or sample are independent (meaning, the measurement taken from one individual is not influenced by the measurement taken from another person).

> **Key Points**
>
> - A concern with the use of parametric tests in a clinical study is whether the data are normally distributed. Most large data sets will tend to be normally or near normally distributed; smaller data sets can be more of a problem. Tests can be done to determine if a data set is normally distributed.
> - Readers of a clinical study should generally not concern themselves with whether or not the data from the treatment groups being compared have equal variances. Refer to Chapter 9 for the definition of variance. Statisticians should best handle situations of unequal variances.
> - The assumption of independent observations/measurements is also not of concern in most clinical studies. Data collected from one patient in a control or treatment group would not usually be affected by the data from another person.

Key Point

Is it acceptable to use multiple t-tests instead of ANOVA to determine which two group comparisons shown in Figure 8.1 (e.g., control versus treatment 1, control versus treatment 2, treatment 1 versus treatment 2) are statistically significant?

No, unless a specific correction is made. The greater the number of individual comparisons, the greater the risk that through chance alone (and not an actual treatment effect) one or more of the comparisons might be found statistically significant (refer to Chapter 9 for a discussion of errors during statistical testing). To help offset this error risk, a correction such as the Bonferroni method can be applied to the use of multiple two-group t-test comparisons. The Bonferroni procedure is acceptable for relatively small numbers of such comparisons. Alternatively, ANOVA is used followed a specific multiple comparison procedure.

A t-test should be used when the means of only two groups are being compared. There are two types of t-tests: *paired* (used for comparisons of paired data) and *unpaired* (used for comparisons of unpaired data). Although ANOVA can be used for a comparison of the means of only two groups (it is the same as an unpaired t-test in this situation), it is usually used for comparisons of means in three or more study groups. There are a number of *ANOVA tests*, including *one-way*, *two-way*, and *repeated measures*, among others (Table 8.1). All of these tests are parametric.

When there are three or more study groups, several individual comparisons are involved with the use of ANOVA (Figure 8.2). When ANOVA is used to analyze the difference in means among three or more groups and finds a significant difference (refer to Chapter 9 for discussion of statistical significance), it provides an overall result that does not indicate which individual group comparison(s) is/are statistically significant. A *multiple comparison test* (also referred to as a "post hoc" test) is used following a significant ANOVA to identify which of the individual two group mean comparisons is/are in fact

Table 8.1 Common analysis of variance (ANOVA) tests

ANOVA test	Used for comparison of means in following situations
One-way	≥3 groups; one independent variable; parallel study design (unpaired data)
Two-way	≥3 groups; two independent variables; parallel study design (unpaired data)
Repeated measures	≥3 groups; one independent variable; cross-over design, time series design, or before (baseline) and after therapy in a parallel design (paired data)
	≥2 groups; one independent variable; parallel design; multiple (repeated) measures taken over time in each study group (unpaired data)

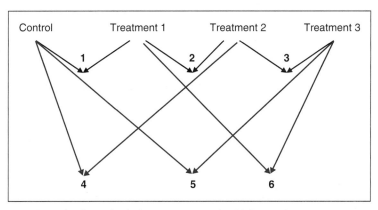

Figure 8.2 Example of six individual comparisons made with four study groups.

statistically significant. Some of the more commonly used multiple comparison tests include Scheffe's test, Tukey's honestly significant difference (HSD) test, Newman–Keuls test, Dunnett test, and the Fisher least significant difference (LSD) test.

Worked example

Example 8.1

A parallel double-blind study is performed to compare the diastolic blood pressures after 12 weeks of therapy in patients randomized to receive enalapril ($n = 68$), lisinopril ($n = 72$), or fosinopril ($n = 65$). Assume the blood pressure readings are normally distributed.

Which of the following statistical tests should best be used to analyze these data: paired t-test, unpaired t-test, one-way ANOVA, two-way ANOVA, or repeated measures ANOVA?

One-way ANOVA. A t-test should not be used here since there are three groups, not two. There is one independent variable (drug therapy) so a two-way ANOVA would not be appropriate. Since the groups are unpaired (parallel design) and the blood pressure is being measured once at the end of 12 weeks, a repeated-measures ANOVA would be inappropriate.

Nonparametric statistical tests

There are several commonly used nonparametric tests in practice. The selection of the test to use depends on whether the data being analyzed are nominal or ordinal level, the number of groups involved, and whether the data are from paired or unpaired samples. The *Chi-square*, *Fisher's exact*, and *McNemar* tests are among those used for comparing between group differences for nominal-level data.

Key Point

A rule of thumb is that at least 80% of the contingency table cells should have an expected frequency of at least 5. In a table with four cells (Example 8.2), 80% of 4 = 3.2, so all cells would need to have an expected frequency ≥5 in order to use the Chi-square test.

There is more than one type of Chi-square test. The *Chi-square test* of proportions is used to compare two or more independent (unpaired) groups of nominal-level (categorical) data involving frequencies or proportions. The data are placed in a *contingency table* with treatments in the rows and outcomes in the columns; each data point in the table constitutes a cell. The Chi-square test compares the values observed in the study with expected values, those that one would expect to see if the treatments had the same effect (i.e., were not different). To use the Chi-square test, the sample size should be large enough so that the expected frequency in each cell is at least 5.

Worked example

Example 8.2

A study compared the effect of amoxicillin therapy to placebo for treating otitis media. A total of 48 children were studied, with 24 children assigned to receive amoxicillin and the other 24 children assigned to receive placebo. Following 1 week of therapy, 20 children in the amoxicillin group were cured compared to 18 children in the placebo group.

Present these data in the form of a contingency table.

Answer:

Therapy	Outcome	
	Cure (number of patients)	No cure (number of patients)
Placebo	18	6
Amoxicllin	20	4

The values of 18, 6, 20, and 4 are the cells in the table. There are four cells in a 2 × 2 table.

If one of the cells in a 2 × 2 contingency table is <5, the *Fisher's exact test* can be used. The Fisher's exact test is used for 2 × 2 comparisons (two independent groups) of nominal-level data when the Chi-square assumption is violated. For comparisons of two groups of nominal-level data when the groups are *not* independent (i.e., for paired data), the *McNemar test* is used instead of the Chi-square test.

The *Mann–Whitney U* (also known as *Wilcoxon rank-sum*), *Wilcoxon signed-rank*, *Kruskal–Wallis*, and *Friedman* tests are among the nonparametric tests used for ordinal data. In addition, these tests can be used

Table 8.2 Parametric tests and nonparametric counterparts

Parametric test (continuous data)	Nonparametric test (ordinal data or continuous data when a parametric assumption is violated)
Unpaired t-test	Mann–Whitney U (Wilcoxon rank-sum)
Paired t-test	Wilcoxon signed-rank
One-way ANOVA	Kruskal–Wallis
Repeated-measures ANOVA	Friedman

ANOVA, analysis of variance.

as alternatives to the t-test or ANOVA for continuous-level data when one or more of the assumptions for parametric tests are not met. The Mann–Whitney U/Wilcoxon rank-sum test is used for comparing two independent (unpaired) groups; the Wilcoxon signed-rank test is used instead when the two groups are paired. The Kruskal–Wallis test is used for comparing three or more independent (unpaired) groups; the Friedman test is used instead when the groups are paired or for multiple (repeated) measures in patients. The nonparametric counterparts to the commonly used parametric tests are shown in Table 8.2.

Worked example

Example 8.3

For each data description on the left, match the statistical test on the right that would best be used for its analysis.

_____ 1. Comparison of the distance walked in 6 minutes between patients receiving either placebo (102 patients) or sildenafil (125 patients) to treat pulmonary hypertension (data not normally distributed)

_____ 2. Percentage of patients receiving placebo (40 patients), drug A (59 patients), or drug B (48 patients) who experienced headache or nausea during therapy (no cell frequency < 5)

_____ 3. Comparison of patients' ratings of joint stiffness (on scale of 1–4) at baseline and after 3 months of therapy with methotrexate

_____ 4. Low-density lipoprotein cholesterol concentrations in patients receiving atorvastatin (120 patients), simvastatin (129 patients), or lovastatin (135 patients) (data normally distributed, variances equal)

a. Paired t-test
b. Mann–Whitney U
c. One-way ANOVA
d. Unpaired t-test
e. Kruskal–Wallis
f. Chi-square
g. Wilcoxon signed-rank
h. Fisher's exact

Answers

1.b. Two unpaired groups; distance walked is continuous-level but assumptions not met for parametric test use.
2.f. Two unpaired groups; percentages (proportions) of patients with headache or nausea are nominal-level; sample size sufficiently large for Chi-square use since no cell has a frequency <5.
3.g. Two paired comparisons – before and after therapy in same patients; ordinal-level ranked data.
4.c. Three unpaired groups; continuous-level data that meet parametric test assumptions.

Correlation

Key Point

A negative r can indicate a correlation as strong as, or stronger than, a positive r; it's just that the direction changes. The following values for r provide a rough estimate of the strength of the correlation:

- strong: -0.5 to -1 (or 0.5 to 1)
- moderate: -0.3 to -0.5 (or 0.3 to 0.5)
- weak: -0.1 to -0.3 (or 0.1 to 0.3)
- none: <-0.1 (or <0.1).

The association between two variables is called a correlation. Often a study will want to determine the extent to which one variable affects another variable. For example, investigators may wish to examine the extent to which diazepam blood concentrations affect a person's reaction time while driving. A *correlation coefficient*, represented as r, is used to quantify the degree and direction of a linear association (correlation) between two variables. The values of r can range in value from -1 to 1 (Figure 8.3):

- 0 indicates no linear association between the variables (i.e., whether the value of one variable increases or decreases has no effect on the value of the other variable).
- 1 (positive or negative) indicates perfect correlation, meaning as one variable changes in value the other variable changes by the same proportion.
- A negative r indicates a negative (inverse) association, meaning as one variable increases in value, the other variable decreases (or the reverse).

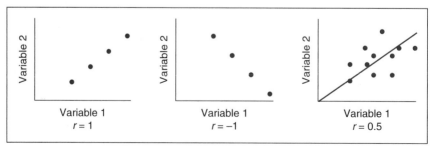

Figure 8.3 Illustration of correlation (r) values.

- A positive *r* indicates a positive association; as one variable increases (or decreases) in value, the other also increases (or decreases) in value.
- The closer the *r* is to 1 (in either direction), the stronger the correlation.

Two common statistical methods are used to calculate a correlation coefficient *r* value: the *Pearson r* (or *Pearson product-moment r*) and the *Spearman rank-order (rho) r*. The *Pearson r* is parametric, meaning that the values for both variables should be continuous-level and normally (or near normally) distributed. The *Spearman rank-order r* is its nonparametric counterpart and is used when one or both variables are either ordinal-level or continuous-level but not normally distributed.

Worked example

Example 8.4

In the previous diazepam example, the investigators report a correlation coefficient of $r = 0.65$ for diazepam concentrations and reaction time (in seconds) at a driving simulator.

How should this correlation be interpreted?

The positive value means that, as diazepam concentration increases, reaction time increases as well. The value of 0.65 indicates a fairly strong correlation since the closer it is to 1, the stronger the correlation.

Key Points

Keep in mind the following about the correlation coefficient *r*:

- *r* values are not directly proportional to each other; for example, $r = 0.8$ for one correlation analysis is not twice as strong as another $r = 0.4$.
- *r* values are on a continuum, with no precise cut-off for which correlation is "strong" and which is "moderate."
- The statistical significance of an *r* value can be calculated. Large sample sizes might find that an *r* value is "statistically significant," meaning not likely a result of chance but due to a real association between variables, but the *r* might still represent a weak correlation.
- r^2 (also called the *coefficient of determination*) is helpful for interpreting the degree of association between two variables. It represents the amount/proportion of variation in one variable that can be explained by the presence of the other variable. In the diazepam example, $r^2 = 0.65^2 = 0.42$. This means that 42% of the variation in reaction time can be explained by knowing the diazepam concentration. The remaining 58% of reaction time variability would be due to other factors unrelated to diazepam concentrations.
- Just because there is a strong correlation between two variables does not mean that one variable caused the other. For example, a strong correlation might be found between walking one's dog and eating healthy foods. This does not mean that walking a dog causes a person to eat healthier foods (or that eating healthier foods causes one to walk a dog). They may be correlated because persons who exercise to a greater extent take better care of their overall health, including healthier eating habits.

Table 8.3 Statistical method selection and use

Scale (level) of measurement of data	Type of study design or experiment			
	Two groups – parallel (unpaired)	Two groups – cross-over or before/after measures in same patients (paired)	Three or more groups – parallel (unpaired)[a]	Multiple measures (3 or more) in ≥2 groups; or ≥3 groups – cross-over or before/after measures in same patients (paired)
Nominal	Chi-square test Fisher's exact test	McNemar test	Chi-square test	
Ordinal[b]	Mann–Whitney U test (Wilcoxon rank-sum)	Wilcoxon signed-rank test	Kruskal–Wallis test[c]	Friedman test
Continuous[d]	Nonpaired t-test	Paired t-test	One-way ANOVA Two-way ANOVA	Repeated-measures ANOVA

[a] Analysis of variance (ANOVA) can be used for two groups but t-test is usually used instead.
[b] Also used for continuous data for which one or more of parametric test assumptions not met (e.g., non-normally distributed, unequal variances between groups).
[c] For only one independent variable.
[d] Assume parametric test assumptions met; if not, use appropriate test described for ordinal data.

Regression

Regression involves predicting the value of an outcome measure based upon the value of an independent (explanatory) variable. Examples of regression analyses include, but are not limited to: (1) simple regression (one continuous-level independent variable; one continuous-level outcome measure); (2) multiple regression (two or more categorical (nominal-level) or continuous independent variables; one continuous-level outcome measure); (3) simple logistic regression (one categorical or continuous independent variable; one categorical outcome measure); and (4) multiple logistic regression (two or more categorical or continuous independent variables; one categorical outcome measure). An example of a multiple logistic regression analysis involves studying the relationship among age, gender, number of drugs taken, number of prescriptions received (a variety of categorical and continuous independent variables) and whether or not a person was more likely to die from prescription drug use (one categorical outcome – alive or dead).

A summary of key points and how to apply the information from this chapter to practice follow.

Key Points

- Table 8.3 summarizes the considerations guiding the selection of commonly used parametric and nonparametric tests.
- A correlation coefficient provides the extent of linear association between two variables; it can only assume a value from −1 to 1 and a negative value indicates an inverse association between variables.

How to apply to practice

When evaluating the statistical analyses used in a clinical study, consider the following:

- The analysis should be appropriate based on the number of groups, whether the data were paired or unpaired, and the level of data involved.
- Parametric tests should be used whenever appropriate.
- t-tests should not be used to analyze three-group comparisons unless a correction such as the Bonferroni correction is employed; ANOVA is the parametric test preferred for three or more group comparisons.
- If a study uses a nonparametric test for continuous-level data, assume that a parametric test assumption, most likely normally distributed data, was violated.
- Look at the actual value of r calculated for a correlation coefficient, regardless of whether it was found to be statistically significant. Weak r values, indicating a lack of meaningful association between variables, could be statistically significant if the study enrolled a large sample size.
- r^2 can provide a good way to interpret correlation coefficients in a study.

Self-assessment questions

For each of the types of data comparisons shown in questions 1–9, indicate which of the following statistical tests would be most appropriate to use for its analysis: paired t-test, unpaired t-test, one-way ANOVA, two-way ANOVA, repeated-measures ANOVA, Chi-square, Fisher's exact, McNemar, Mann–Whitney U, Wilcoxon signed-rank, Kruskal–Wallis, Friedman, Pearson r, or Spearman rank-order r.

Question 1
Number of chemotherapy patients experiencing moderate to severe vomiting after receiving a control antiemetic, ondansetron (18 of 58 patients; 31.0%), or a new antiemetic drug (14 of 60 patients; 23.3%) in a randomized study.

Question 2
Comparison of serum hemoglobin concentrations measured each week for a total of 4 weeks in a randomized, parallel study of patients taking a new IV iron product, an older IV iron product, or placebo (assume data normally distributed, variances equal).

Question 3
Association (linear) between clozapine drug concentrations and overall symptom scores (0 = no symptoms; 1 = mild symptoms; 2 = moderate symptoms; 3 = severe symptoms) in a study of patients receiving clozapine therapy.

Question 4
Comparison of body weights at the end of therapy in 68 patients who were randomized to receive one of three antipsychotic drugs for 24 weeks (data normally distributed, variances equal).

Question 5
Change in body weight from baseline (prior to initiating therapy) to the end of therapy following treatment with an antipsychotic drug for 24 weeks (data normally distributed, variances equal).

Question 6
Comparison of serum glucose concentrations in patients after receiving dexamethasone or placebo eye drops in a cross-over study (data not normally distributed).

Question 7
Comparison of the frequency of dizziness development in patients receiving candesartan or amlodipine to treat hypertension, in a cross-over study of 75 patients.

Question 8
Comparison of visual analog scale (VAS) pain scores at the end of therapy in patients with temporomandibular joint pain randomized to receive either physical therapy alone (18 patients) or physical therapy plus analgesics (19 patients) (data not normally distributed).

Question 9
Comparison of the bone marrow density (BMD) scores in postmenopausal women randomized to receive one of two doses of ibandronate or placebo for the treatment of low bone density (BMD scores normally distributed, variances equal). In addition to the effect from the treatment received, the statistical analysis used also accounted for the effect of the patients' baseline BMD score on the outcome measure.

A study examines the association between quality of life in arthritis, measured using a scale of from 3 (excellent) to 0 (poor), and the extent of joint erosion (measured in millimeters). The study reports an $r = -0.57$ (Spearman rank r) for this association. Answer questions 10–13 about the r value reported.

Question 10
What does the negative value of r indicate?

Question 11
Was it appropriate to use the Spearman rank r to determine the correlation?

Question 12
Is the correlation reported considered to be strong?

Question 13
How much of the variability in the patients' quality of life scores can be explained by variations in the extent of joint erosion?

Question 14
Investigators performed a study to compare the efficacy of two strengths of phenytoin topical cream, 0.5% and 1%, to cream vehicle alone (placebo) to promote wound healing following the excision of minor skin lesions. Patients were randomized to receive one of the three treatments applied daily for 3 weeks. The change in wound size from baseline to the end of therapy was compared among groups. The study reported that they used ANOVA with the post hoc least significant differences (LSD) method for the statistical analyses. Was it appropriate for the LSD method to be used here?

Summary

Research studies generate data from outcome measures, and statistical analyses are used to "make sense" of the data. Statistical analyses can determine the likelihood that the treatment being studied produced the results seen, can show whether there are associations among the variables in the study, and can help the reader predict the results that might be expected to occur in the population outside the study. This chapter reviewed several of the common statistical tests used to analyze study data, including the requirements for use of the various tests.

9

Evaluating clinical studies – step 5: results, interpretation, and conclusions

Learning objectives
Upon completion of this chapter, you should be able to:
- describe the measures of central tendency of data, including mean, median, and mode
- discuss and interpret important measures of the variability of study data, including standard deviation, standard error of the mean, range, and interquartile range, and define the term variance
- describe the process of statistical inference, including use of confidence intervals, the importance of statistical power and the factors affecting it, and the interpretation of *P* values
- discuss the potential errors in statistical testing and in interpreting study findings: alpha and type I error and beta and type II error
- describe, calculate, and interpret important measures of risk, risk reduction, and clinical utility, including odds ratio, relative risk, relative risk reduction, absolute risk reduction, and number needed to treat
- discuss the importance and impact of patient drop-outs on the clinical applicability of a study's findings and the advantages/disadvantages of the methods used to handle missing data
- discuss the key points that should be included in a study's discussion and conclusions.

Introduction

Interpreting a study's data and the significance of the results are critical aspects of deciding whether, and how, the findings should be applied to clinical practice. This chapter focuses on the study results (data) and conclusions made from them. Determining how most study patients responded to therapy (measures of central tendency) and how widely individual responses are spread around that central point (measures of variability) are addressed. Hypothesis testing is discussed along with statistical inference, the process by which the study sample data are used to make conclusions about the underlying population. There is always the possibility for error in the conclusions made with hypothesis testing and statistical inference, so this chapter will include: *P* values, alpha, and type I errors; power, beta, and type II errors; and statistical versus clinical significance. We will also review how the method used for handling patient drop-outs can affect the likelihood of error.

This chapter will describe how to apply the study results to clinical practice through the use of measures of risk, risk reduction, and clinical utility, such as the odds ratio, relative risk, relative risk reduction, absolute risk reduction, number needed to treat, and number needed to harm. Finally, considerations when reading a study's discussion section and conclusions will be reviewed. For more detailed information and additional examples of the concepts reviewed in this chapter, the reader is referred to *How to Report Statistics in Medicine* (Lang and Secic 2006), *Intuitive Biostatistics* (Motulsky 2010), *Introduction to Research. Understanding and Applying Multiple Strategies* (DePoy and Gitlin 2011), and Abate (2012).

Measures of central tendency

When clinicians read a study's results, they would like to know how the "usual" or "typical" patient in the study responded to therapy (i.e., the *central tendency* of the data) and not just look at a listing of individual patient values. Three measures of central tendency that studies will report for their data are the *mean, median,* and *mode*.

The *mean* is calculated as the arithmetic average of a data set. It is determined by adding each of the values in the data set and dividing the sum by the total number of values (n). For example, if serum potassium values (in mEq/L) for 10 patients in a study were: 4.1, 3.1, 5.2, 3.7, 5.1, 3.2, 4.8, 4.3, 3.9, and 5.1, the mean would be calculated as:

> **Key Point**
>
> Although investigators will often report means for ordinal data, use caution when interpreting these results.

$$\frac{4.1 + 3.1 + 5.2 + 3.7 + 5.1 + 3.2 + 4.8 + 4.3 + 3.9 + 5.1}{10} = 4.25$$

Means are commonly reported in studies and provide a useful estimate of the central tendency (clustering) of continuous-level data. Means have also been used for ordinal-level data, particularly for questionnaire or survey results that use rankings such as 5 = excellent/strongly agree; 4 = very good/agree; 3 = average/neutral; 2 = poor/disagree; 1 = very poor/strongly disagree. When a mean is reported for ordinal data, keep in mind that the distance between numbers, e.g., between 1 and 2, or 4 and 5, is not necessarily equal. As a result, it is difficult to interpret what a decimal really signifies with mean values such as 3.7 versus 3.4 for ordinal-level data.

A problem with the mean is that it can misrepresent the central clustering when reported for data that have outliers, defined as a small number of very large or very small data points compared to the rest of the data set. For example, suppose a study measured 10 estrogen concentrations in women and reported the following values (in pg/mL): 28, 29, 30, 30, 29, 28, 28, 30, 30, and 259. Due to the outlier, the mean of these data is 52.1. This misrepresents (is much higher than) the clustering of the rest of the data around the value of 29.

> **Key Point**
>
> Look for the median as a better indicator of the central clustering of study data for non-normally distributed (skewed) data. The mean is best used for normally or near normally distributed data. How can you tell if data are non-normally distributed? The mean and median have the same value for normally distributed data (think of a bell-shaped curve). If the median and mean differ substantially, the data are skewed.

The *median* is the midpoint of a rank-ordered listing of all data points (also referred to as the 50th percentile). With an odd number of values, the median is the middle value. With an even number of values, the median is the average of the two middle values. For the estrogen example above with 10 concentrations, the median is the average of the two middle values, 29 and 30, which would equal 29.5: 28, 28, 28, 29, 29, 30, 30, 30, 30, 259. Note the advantage of the median is that it is not affected by a small number of outlying data points, in this case the value of 259. Thus, it better represents the central tendency of data that has one or more outliers.

The *mode* is the most frequently occurring value in a data set. Although in theory it could be reported for continuous-level data, it makes the most sense when reported for data consisting of whole numbers. Why? If a study reports continuous data measured to one decimal place and all the values are close but slightly different, by definition there would be no mode. Only the mode can be used for nominal data. For example, suppose a study reports systolic blood pressures in patients using the following categories: <120 mmHg (23 patients); 120–149 mmHg (27 patients); ≥150 mmHg (18 patients). The mode is 120–149 mmHg (the category with the most (27) patients). A mode can be reported for ordinal data, e.g., if patients rank their satisfaction with therapy on a scale of 0–4 and most patients indicate a "3," 3 would be the mode.

Table 9.1 summarizes the measures of central tendency used for each scale/level of measurement.

Worked example

Example 9.1

A study evaluating the efficacy of a herbal Chinese tea extract for treating hyperlipidemia in 12 diabetic patients reported that the serum cholesterol concentrations following 8 weeks of therapy were: mean = 220 mg/dL; median = 175 mg/dL.

Which value appears to provide a better estimate of the central clustering (tendency) of these data?

The median would provide the better estimate of central tendency in this example because the data appear skewed. If the data were normally distributed, the mean and median would be equal or close in value. Since they are substantially different, the data are likely skewed. With skewed data, outliers do not affect the median but will affect the mean.

Measures of variability (spread or dispersion)

In addition to knowing how a typical patient responded in a study (i.e., central tendency of data), clinicians would like to see how spread out the data (patient responses) were. Did *every* patient respond exactly the same as the mean or median value, or were many higher or lower (better or worse) responses present? Measures of variability tell the reader the spread or dispersion of the individual study data points. Measures of variability commonly reported in clinical trials include the *range, interquartile range* (*IQR*), *standard deviation* (*SD*), and *standard error of the mean* (*SE* or *SEM*).

The *range* is the spread (difference) between the highest and lowest value in a data set, although most authors report it as the lowest and highest values. A related, more useful measure that provides a better indication of where the values fall within the range is the *IQR*. The IQR includes those individual values that fall between the 25th and 75th percentiles (which comprise 50% of the values); authors usually report IQR as the values for the 25th and 75th percentiles. For example, a study reported adherence to therapy as the percentage of doses taken and listed the IQR as 55–96%. This IQR indicates that 50% of the adherence values were between 55% and

Table 9.1 Use of the measures of central tendency

Measure	Scale/level of measurement
Mean	Ordinal data[a] Continuous data[b]
Median	Ordinal data Continuous data
Mode	Nominal data Ordinal data Continuous data[c]

[a] Use caution when interpreting a mean value for ordinal-level data.
[b] Can misrepresent central tendency for skewed data; median preferred for skewed data.
[c] Not usually helpful here.

96% (the 25th and 75th percentiles); thus, 25% of the values were ≤55% and 25% were ≥96%. Note that outliers would be excluded from the IQR.

The *variance* is not usually reported in clinical trials but you may see this term. It is essentially looking at the differences between each individual data point and the mean value. It is calculated as the average of the difference between each individual data point and the mean value, in which each difference is also squared.

In contrast to the infrequently used variance, the square root of the variance is called the *standard deviation*. The SD is one of the most commonly reported measures of variability in clinical studies and gives the spread of the individual study values around the mean value for that measure. For a normal distribution of data, ~68% of the individual values will be found within ±1 SD, ~95% of the values will be within ±2 SD and ~99% of the values will be within ±3 SD of the mean value.

Key Point

- IQR is usually reported with the median; these measures are preferred for skewed (non-normally distributed) data. The smaller the IQR, the less spread out the individual values are.
- SD is usually reported with the mean; these measures are preferred for normally or near-normally distributed data. The smaller the SD, the closer the individual values lie to the mean.

Studies will report a mean and SD as: mean (SD) or mean (± SD). Although it doesn't say "1 SD," the number listed would be for 1 SD. For example, suppose a study reported a mean (SD) age in years for their patients as 54.5 (8.7). In this case, approximately 68% of the patients' ages (assuming a normal or near-normal age distribution) were between 45.8 years (54.5 − 8.7) and 63.2 years (54.5 + 8.7). Note that ~34% of the ages would be found on each side of the mean for 1 SD. About 95% of the patients' ages would be between 2 SD, or between 37.1 years (54.5 − 17.4) and 71.9 years (54.5 + 17.4). Remember that the percentages indicated above apply to normally or near normally distributed data. If the data are skewed, the SD *cannot* be interpreted in this way.

Worked example

Example 9.2

Two studies examined whether counseling diabetic patients about their medications increased blood glucose control. The patients' mean (SD) fasting blood glucose concentrations following counseling were: study 1, 160 (31) mg%; study 2, 158 (45) mg%.

Which study, 1 or 2, reported greater variability in individual patient responses following therapy?

Study 2 shows greater variability of individual patient responses since the reported SD is larger. Assuming normally or near normally distributed glucose concentrations, ~68% of the individual patients' concentrations in study 1 were within 129–191 mg%, compared to 113–203 mg% for study 2.

Worked example

Example 9.3

Investigators determined the efficacy of a new potassium supplement on serum potassium concentrations in 20 patients taking a diuretic who had hypokalemia. Following 2 weeks of therapy with the new supplement, the serum potassium concentration was *4.1 + 0.3 mg%* (mean ± SD).

Interpret the meaning of the italicized statement.

The mean (arithmetic average) of the 20 individual patients' potassium levels was 4.1 mg%, which provides an indication of the "central tendency" or "typical" value for the data. The SD is 0.3 mg%, which indicates that ~68% of the individual patients' potassium concentrations were within 3.8–4.4 mg% (calculated from the mean of 4.1 − 0.3 and 4.1 + 0.3). This assumes that the data are normally or near normally distributed.

Standard error of the mean is frequently reported in clinical trials, although in most instances it should *not* be reported. SEM provides an estimate of the variability or precision of the *study sample* mean value for an outcome measure around the unknown *population* mean value. In other words, if a study were repeated multiple times, selecting a different patient sample each time from the same target population, each study would probably report a somewhat different mean value for the outcome since the patients would be different in each study. If the mean values from these individual studies were plotted around the actual (but unknown) population mean, you would find a distribution of study mean values. Assuming that each individual study enrolled a large sample size, by definition about 68% of the study mean values would fall ±1 SEM from the actual population mean value. Note two important things here: (1) SEM is *not* referring to variability or spread of any of the individual data *within* the study; and (2) SEM is rather theoretical and generally not of much use to clinicians.

The SEM is calculated from the SD as follows: SEM = $SD/\sqrt{n}$. Also note that the calculated SEM is *always* a smaller number than the SD. *The problem with SEM is that clinical studies will sometimes report SEM instead of SD simply because SEM is a smaller number.* Clinicians quickly reading a study without paying much attention to the abbreviations might automatically assume they are seeing a SD with the mean value instead of the SEM.

Key Points

Use the SD to determine the spread of individual patient responses in a study. If a clinical study only reports a SEM value, *always calculate the SD* because SD is more useful for clinicians.

Why would investigators report SEM instead of SD?

Most likely, to make the spread of the individual study responses appear less (closer around the mean) than it actually is. If a study reports a mean ± SEM, consider whether there are any potential conflicts of interest. For example, a pharmaceutical manufacturer that performed the study would want patients to respond to the drug in as consistent and uniform a manner as possible (i.e., as illustrated by a small SD). A large SD means there is a lot of individual patient variability in drug response (i.e., individual responses could be much larger or smaller than the mean value). Thus, some patients could have a great response to therapy while others have little or no effect, or perhaps are even worse. To clinicians, this indicates that their patients could respond unpredictably to the drug in practice. As a result, they might hesitate to prescribe the drug (not what the manufacturer would desire).

Is there an important use for SEM?

Yes, SEM is used to calculate the confidence interval (CI). As a clinician, use the CI instead to estimate the underlying population value (see CI discussion).

Worked example

Example 9.4

A study examined the efficacy of a new amlodipine + diuretic combination for hypertension treatment. A total of 200 patients were in the 16-week study. At the end of week 16, the change from baseline in mean (SEM) systolic/diastolic blood pressure was −18.7 (0.8)/−10.9 (0.5) mmHg.

Was it appropriate for the SEM to be reported? Would SD be better? If so, what would the SD be for this finding?

No, SEM should not have been used here. It does not provide the variability (spread, dispersion) of the *study* data. The SD is more useful for clinicians, who would like to know how widely spread the individual patients' blood pressure responses were to the therapy. In many studies that inappropriately report the SEM, there is a potential conflict of interest (e.g., manufacturer funding the study). The investigators may want the study variability to look small to clinicians who might quickly scan the results presented.

You should calculate the SD any time SEM is reported with the mean in a clinical study. In this example, the SD can be calculated from the following equation: $SEM = SD/\sqrt{n}$. Thus, $SD = SEM \times \sqrt{n}$. The SD for the systolic blood pressure would be $0.8 \times \sqrt{200} = 0.8 \times 14.1 = 11.5$. Notice what a difference it makes to see −18.7 (11.5) versus the −18.7 (0.8) reported. The SD for the diastolic blood pressure = $0.5 \times \sqrt{200} = 7.05$.

Statistical inference

Once a study has collected data from the outcome measures, important questions arise. What do these data tell us about the population of interest as a whole? Do these data support the study's hypothesis? *Statistical inference* is the process used to draw conclusions about the underlying population from the evidence (data) obtained in the study sample.

We covered the measures of central tendency and variability commonly used to describe a study's data. Is there a measure that can now help us apply a study's result to the population? The *CI* can do that.

Confidence interval

Suppose the US presidential election was nearing and the following was reported: "A telephone poll of 2014 voting citizens representing Americans from across the USA showed candidate A had a 5% lead over candidate B, with 54% favoring candidate A and 49% favoring candidate B (i.e., 5% difference)." Can we say with *100% certainty* that in the actual presidential election, candidate A would beat candidate B by *exactly a 5% difference*? Now, suppose we take the election example and substitute a typical clinical trial (let's simply replace the election wording with clinical trial information as follows):

Suppose ~~a U.S. presidential election was nearing~~ a study compared the efficacy of two drugs and the following was reported: "~~A telephone poll of~~ 2014 ~~voting citizens representing Americans from across the U.S.~~ patients were randomized to receive either drug A or drug B. The results showed that ~~Candidate~~ drug A had 5% greater efficacy than ~~Candidate~~ drug B, with efficacy reported in 54% of patients taking drug A and 49% of patients taking drug B (i.e., 5% difference)." Can we say with *100% certainty* that in the actual patient population outside the study, drug A would have greater efficacy than drug B by *exactly a 5% difference*?

The answer for both examples above is, "No." Why not? Because in the first example, even if the telephone poll truly represented the diversity of all voters, the poll still only looked at a *sample* of voters. The same is true for the drug efficacy study: only a *sampling* of patients from the population of interest participated in the study. In both examples, wouldn't it be nice to know, let's say with 95% certainty, what the election result or drug efficacy difference would likely be in the *population* outside the study? Is there a measure that could provide this information? Yes, the CI is such a measure.

The CI *provides the likelihood of what the population value would be for a specified study finding*. The CI is usually stated as a 95% CI (occasionally a 90% or 99% CI is used) followed by a range of values likely to include the actual population value. A 95% CI is saying that one is 95% confident that the range of values provided will contain the population value for the specified outcome measure.

There are actually two ways to interpret a CI in a study. Suppose a study states the following: "The clinical cure rate was 91.8% (89 of 97 patients) for levofloxacin compared with a cure rate of 82.4% (84 of 102 patients) for ciprofloxacin (9.4% difference between drugs; 95% CI = 2.1–16.8%)." One way to interpret this CI is: there is 95% confidence that the range of 2.1–16.8% will contain the population difference in clinical cure rates between levofloxacin and ciprofloxacin. Another way to interpret this CI is: if this study were repeated *many, many* times (not just once or twice), each time with a different sample of patients taken from the population, in 95%

of those studies the difference in clinical cure rates between levofloxacin and ciprofloxacin would be between 2.1% and 16.8%.

Key Points

Will we ever know *exactly* what the actual population value is?

No. In clinical situations, it is virtually impossible to study the *entire* population. This is why the CI is useful to clinicians. It gives readers the range of values likely (at a specified level of confidence) to contain the actual population value for that measure.

Who exactly is in the *population*? People like you and me?

The population consists of those individuals who meet the inclusion and exclusion criteria specified in the study, not everyone in the world (you and I would be in the population if we met the study criteria!).

What types of data can a CI be reported for?

A CI can be calculated for nominal- or continuous-level data. In a clinical study it can be determined for efficacy rates or other outcome measures within a group, as well as for differences between groups.

How should a CI with a negative value be interpreted?

Remember that the CI provides the range of values likely to contain the actual (unknown) population value. Negative numbers in the CI simply mean that the population value might be a negative number for whatever that outcome measure was. Let's take the election/drug example from earlier and add the following CI: "The results showed that drug A had greater efficacy than drug B, with efficacy reported in 54% of patients taking drug A and 49% of patients taking drug B (5% difference, 95% CI of −3% to 13%)." The CI is given for the difference in efficacy between drugs A and B. The 5% difference indicates that the drug B efficacy was subtracted from the drug A efficacy (54% − 49% = 5% difference). The negative CI value of −3% indicates that drug A efficacy could be 3% *less* than drug B efficacy, a value of 0 indicates that both drugs have the same efficacy (one subtracted from the other would give a value of 0), and positive values indicate that drug A has greater efficacy. Thus, this CI would be interpreted as: 95% confidence that the population value for the difference in efficacy between drug A and drug B could range from 3% less efficacy with drug A to as much as 13% greater efficacy with drug A compared to drug B.

How should one interpret a very wide CI versus a very narrow CI?

A very wide CI means that the actual population value could assume any of the values in that range. For example, suppose a drug was found to be efficacious in 58% of patients, with a 95% CI = 18–98%. This means that the actual drug efficacy in the population could be as low as 18% or as great as 98%. Given the extreme variability, it is difficult to make any definitive conclusion about the clinical efficacy of this drug in practice.

What factors affect the width of the CI range?

Factors that affect the CI width include:

- Level of confidence selected: 90% CI   95% CI   99% CI
                                 narrower   wider   widest

    Since there is less confidence with a 90% CI that the actual population value is contained within that range of values, the range will be narrower. Conversely, if one wishes to be very confident that the actual population value will be in the CI range of values, the 99% CI will provide the widest range of values.

- Sample size: as the sample size increases in a study, the better the population will be represented by the study sample and the narrower the CI range of values will be.
- SD of study sample: the greater the SD (individual variability of patient values) in a study, the more difficult it will be to predict the population value and the wider the resulting CI.

Worked example

Example 9.5

A study compared the efficacy of pine bark extract (64 patients) with placebo (56 patients) for the treatment of hypertension. Patients receiving pine bark extract had a mean decrease in diastolic blood pressure from baseline to the end of therapy of 3.1 mmHg (95% CI = 1.0–5.2 mmHg).

a. Interpret the meaning of this 95% CI

It means that one is 95% confident that the mean decrease in diastolic blood pressure with pine bark extract when used in the population (i.e., those individuals outside the study who meet the study's inclusion/exclusion criteria) could be as small as 1.0 mmHg or as large as 5.2 mmHg. Since most of the values in this CI range are small, clinicians can conclude that the efficacy of pine bark extract in most of the population will probably not be of clinical importance.

b. Suppose this study had reported a 90% CI instead. Compared to the 95% CI, the 90% CI would be (circle the correct answer):

   The same width   Wider than the 95% CI   Narrower than the 95% CI

A 90% CI would be narrower than the 95% CI (all else being equal). Since one is less confident that the population value would be in a 90% CI, the range of values will be smaller than the 95% CI.

c. Suppose this study enrolled 200 patients in the pine bark extract group instead of 64 patients. How would this increased number of patients affect the width of the 95% CI reported?

The CI would be narrower with 200 patients because the greater the number of patients a study enrolls, the better the study sample will represent the underlying population. Thus, the CI range likely to contain the population value will be more precise (narrower).

Hypothesis testing

Statistical inference incorporates *hypothesis testing*, the process of determining whether or not the data gathered support the study's hypothesis. In a clinical study examining therapy efficacy, hypothesis testing is used to determine whether it was the therapy or some other factor, such as chance, that was likely responsible for any effects observed.

Hypothesis testing is generally used in a study either to reject or fail to reject the null hypothesis, based upon the results of statistical testing. In Chapter 6 we reviewed the *null hypothesis*. In a controlled experimental clinical trial, the null hypothesis assumes that there is no difference between the treatments studied (or no difference between comparisons made before and after therapy within the same study group). Thus, statistical hypothesis testing is used either to reject, or fail to reject (accept), the null hypothesis. If not clearly stated, the null hypothesis is determined from the study's objective.

There are several important concepts in hypothesis testing: probability (P) values, type I error, alpha (α), type II error, beta (β), statistical power, and the factors influencing power. The following discussion will review each of these and describe how they interrelate. For more detailed information, the reader should refer to *What is a P-Value Anyway?* (Vickers 2010).

P values

> **Key Point**
>
> Suppose a study measures the difference in glucose concentrations between a drug and placebo and reports a *P* value of 0.00001 for that difference.
>
> **Can a very small P value prove that the drug caused the difference in glucose concentrations seen?**
>
> No, a *P* value provides the likelihood that chance was responsible for the effect observed and not the treatment(s) studied. The *P* value in this example indicates that the likelihood of chance being responsible for the difference seen, or that the null hypothesis was true, is only 1/100 000 (0.00001). While this is certainly a small number, there is still is a very slight possibility that chance was involved.

The statistical tests covered in Chapter 8 are used to generate a probability or *P value*, which is used to test the null hypothesis. The *P* value provides the likelihood (probability) that the study results seen would have occurred if the null hypothesis were true. Another way to think of the *P* value is that it gives the probability that the difference observed in the study was simply due to chance and not from the treatment(s). Thus, if the *P* value is very small, it is very unlikely that the null hypothesis is true, and there is also a very small likelihood that the difference observed was a chance occurrence. As a result, the treatment is likely responsible for the difference seen and the null hypothesis should be rejected. The larger the *P* value, the greater the likelihood that the difference observed was a chance finding and not from the treatment(s); thus, the greater the likelihood that the null hypothesis is true.

To apply a study's findings to patients, clinicians must decide whether to accept or reject the null hypothesis in a study. However, *P* values are on a continuum that can range between 0 and 1. Is there a commonly accepted "cut-off" for the *P* value below which the null hypothesis should be rejected and above which it should be accepted? Yes, a value of 0.05 (referred to as the level of significance) is generally used for this cut-off in clinical studies. *When $P < 0.05$, it is concluded that the risk of the null hypothesis being true (any difference is due to chance alone) is acceptably small (less than 5/100 or 1/20), so the finding is termed "statistically significant."* When $P \geq 0.05$, the likelihood that the null hypothesis is true is considered unacceptably large, so the finding is *not* statistically significant.

Key Point

The cut-off for statistical significance is usually set at 0.05.

When $P < 0.05$ the following are concluded:
- The finding is statistically significant.
- The null hypothesis is not true (i.e., reject it).
- A treatment effect was found; the difference seen in an outcome measure was not likely a chance finding.

When $P \geq 0.05$ the following are concluded:
- The finding is not statistically significant.
- The null hypothesis is true (i.e., do not reject it).
- A treatment effect was not found; any difference seen in an outcome measure likely represented a chance occurrence.

Other important considerations:
- If a study does not find a statistically significant difference between treatments, it does not automatically mean that the treatments are the *same*. Conclude only that the study *did not find* a significant difference.
- P values do *not* indicate the clinical importance of a difference found; they simply provide the likelihood that the difference was a chance (nontreatment-related) occurrence. Thus, if one P value in a study is 0.02 and another P value in that study is 0.002, it does not mean that the smaller P value is 10 times "better" than the other. It just means that the likelihood that chance might be responsible for the effect seen is 10 times less with $P = 0.002$ compared to $P = 0.02$.

Worked example

Example 9.6

A study compared the efficacy of oral mesalazine ($n = 28$ patients) with topical mesalazine ($n = 31$ patients) for the treatment of distal ulcerative colitis. Following 2 weeks of therapy with either agent, the clinical response rate was 43% with oral mesalazine versus 58% with topical mesalazine ($P = 0.003$).

Is this therapy difference statistically significant?

Yes. $P = 0.003$ indicates that the risk that the null hypothesis is true, or that the difference in clinical response rates is a chance finding and not a real treatment effect, is only 3/1000. Since this P value is less than 0.05, the difference in response rates between topical and oral mesalazine in this study is statistically significant.

P values provide a *guide* for interpreting whether a study's findings were likely due to the therapy or resulted from chance. As stated earlier, chance could still be responsible for a study's finding with a $P < 0.05$. The reverse is also possible: the treatment could still be responsible for differences seen with a $P \geq 0.05$.

Suppose that the conclusion made by looking at the P value is *wrong*, such as: (1) a finding is concluded to be statistically significant and resulting from the treatment given when the difference seen was really due to chance; or (2) a finding is concluded *not* to be statistically significant when the treatment actually caused the difference seen (i.e., a real treatment effect was missed). These two types of erroneous conclusions are termed *type I error* and *type II error*, respectively.

Type I error

Key Point

- The greater the number of statistical comparisons made in a study, the greater the chance that at least one of them will be statistically significant due to chance alone (type I error). For example, if 100 comparisons are made with a $P = 0.03$, by definition three of the comparisons might be found to be statistically significant due to chance alone. Think of running through a golf course holding a metal golf club during a thunderstorm. The more times you do this, the more likely it is that you will be hit by lightning purely by chance. You may never be hit by lightning, but the chance of it increases with each dash through the golf course.
- Be cautious when interpreting the findings from studies that perform many comparisons. Statistical procedures can be used to reduce the risk of type I error from multiple comparisons and should be performed when needed.

A *type I error* is possible when $P < 0.05$. It occurs when it is *wrongly* concluded that a therapy (e.g., drug) produced a significant effect, but the effect seen was really just a chance finding. Type I error can also be defined as rejecting the null hypothesis when it is really true (or a *false-positive* effect).

By definition, *alpha* (α) is the probability of a type I error. Alpha (also known as the level of significance) is usually set at 0.05, as indicated earlier. The P value calculated from a statistical test is then used to determine statistical significance based on whether it is above or below the specified alpha level.

Suppose a study comparing two antihypertensive drugs finds a difference of 8 mmHg in diastolic blood pressure between treatments ($P = 0.03$). Since $P < 0.05$, the difference between treatments is:

- statistically significant
- *unlikely* due to chance (there is a 3/100 likelihood that chance was responsible)
- concluded to represent a real treatment effect.

Is there any way to tell definitively if a type I error occurred with this conclusion? No, we will never know if the blood pressure difference was in fact a chance finding and a type I error was made. All we can say is that the risk of type I error is acceptably low when $P < 0.05$ and the numerical value of P provides an estimate of the type I error risk.

Type II error

> ### Key Points
>
> - Too small a sample size is usually the main reason for inadequate power in a study. Prior to beginning a study, the investigators should calculate the number of patients needed to have a power of at least 80%. The power calculation should be reported for the reader.
> - The best way to increase the statistical power of a study is to increase the number of study patients (i.e., sample size).
> - Check the effect size used for the power calculation to see if it is a reasonable value to use to for a clinically significant effect. Sometimes investigators will use too large an effect size in their power calculation just to make sure that the resulting power is at least 80%. If this happens, realize that the study might have insufficient power to detect smaller but still clinically important differences as statistically significant, thereby increasing the type II error risk.
> - Power is usually calculated for only a specific outcome measure: check to see what measure the power applies to.
> - If a study does not report any power, the risk of type II error for nonstatistically significant findings is unknown and this risk should be considered.
>
> Refer to Table 9.2 for a summary of P value interpretation and types of error.

A *type II error* is possible when $P \geq 0.05$. Type II error occurs when it is *wrongly* concluded that there was no treatment effect or that any difference seen in an outcome measure was due to chance and *not* from the therapy. We fail to reject (accept) the null hypothesis when it is false and should be rejected. Type II error can also be thought of as missing a real treatment effect because the study failed to find a statistically significant difference (i.e., *false negative*). By definition, *beta* (β) is the probability of a type II error.

How can one determine the likelihood of type II error in a study? *Statistical power* helps us determine this. In a clinical study, power is the ability to identify a difference in an outcome measure as statistically significant if an actual treatment effect is present. We appropriately reject the null hypothesis when it is false. Thus, another way of thinking of power is that it is the likelihood of *not* making a type II error. Mathematically, *power = 1 − beta*.

What is the acceptable cut-off value for beta, and power, in a study to minimize the likelihood of a type II error? *A power of ≥80% is desired.* As power increases, the likelihood of type II error decreases. Thus, the greater the power, the better. Since power = 1 − beta, if the desired power is 80% (0.8) or greater, *the desired beta is 0.2 or less.*

How is power calculated? There are equations and formulas available to calculate power, which is beyond the scope of this discussion. However, it is important for clinicians to understand *four main factors affecting*

Table 9.2 Summary of *P* values and types of error

Study finding *P* value and conclusion	If conclusion was in error	
	Type of error if conclusion wrong	How to determine likelihood of error occurring
P < 0.05 Conclusion: Difference found *is statistically significant* and was likely to have resulted from the therapy; not likely due to chance alone; reject null hypothesis	Type I error: Chance was actually responsible for the effect seen; not a real treatment effect (false positive)	Value of *P* gives possibility of a type I error; no way really to know if a type I error occurred
P ≥ 0.05 Conclusion: Difference is *not statistically significant*; likely to be a chance finding and not an actual treatment effect; accept null hypothesis	Type II error: Treatment actually responsible for difference; a real treatment effect was missed (false negative)	β gives possibility of a type II error; power = 1 − β so power gives likelihood of *not* making a type II error

the calculation of power and how they affect power. These factors are summarized below:

1. *Sample size* (the number of patients in the study). In Chapter 7 we discussed sample size and indicated that it should ideally be calculated prior to beginning the study to ensure that sufficient power is present.
   - As sample size increases, power increases.
   - As sample size decreases, power decreases.
   - This is the easiest factor to adjust to increase a study's power.
2. *Effect size* – the size of the difference in the outcome measure that, if present, one would like to identify as statistically significant.
   - For a given number of study patients, as the effect size used in the power calculation increases, the calculated power increases. As the effect size used in the power calculation decreases, the power decreases. For example, in a study of 100 patients, it would be easier to detect a treatment difference (effect size) of 15 mmHg in diastolic blood pressure than a 5 mmHg diastolic blood pressure treatment difference. The risk of missing the larger difference as statistically significant (i.e., making a type II error) would be less than with the small difference; thus, the power would be greater for detecting the larger difference.
   - If a very small effect size is chosen for the power calculation, more patients will need to be included in the study to have adequate statistical power compared to a large effect size.

- The effect size used in a power calculation should usually be the smallest difference that the investigators consider to be clinically important.
3. *Alpha* – risk of type I error. An α = 0.05 is usually selected as the acceptable cut-off for type I error in the power calculation.
4. *Variability of the outcome measure* in the population (estimated by the SD).
    - This refers to the inherent variability of the outcome measure in the population. For example, if serum potassium is the outcome measure, what is the normal SD (variability, spread of data) of serum potassium measurements in a large group of subjects?
    - The value for the SD used in a study's power calculation is identified from previous studies involving that outcome measure. (The investigators have no control over this value.)
    - Note: If the baseline SD in the underlying population is large, a larger number of patients will need to be enrolled in the study in order to identify a significant treatment effect.

Worked example

Example 9.7

Complete the following sentences by filling in the blanks with the correct words:

1. When $P < 0.05$, a _____ (type I, type II) error is possible. With this type of error, one would _____ (reject, fail to reject) the null hypothesis when it is really _____ (true, false). The likelihood of this type of error is referred to as _____ (alpha, beta). This can also be thought of as a false- _____ (positive, negative) finding.
2. When $P \geq 0.05$, a _____ (type I, type II) error is possible. With this type of error, one would _____ (reject, fail to reject) the null hypothesis when it is really _____ (true, false). The likelihood of this type of error is referred to as _____ (alpha, beta). This can also be thought of as a false- _____ (positive, negative) finding.

Answers

1. type I, reject, true, alpha, positive
2. type II, fail to reject, false, beta, negative

Worked example

Example 9.8

Investigators compared a new drug to placebo for the prevention of headaches. The power of the study was not reported. A total of 30 patients who experienced frequent headaches were enrolled, with 15 patients assigned to receive placebo and 15 patients assigned to receive the new drug. Following 10 weeks of therapy, there was a 40% reduction in headache frequency with the new drug compared to placebo ($P = 0.07$). It was concluded that the new drug did not appear to be superior to placebo for headache prophylaxis.

Was the reduction in headache frequency statistically significant? Is a type II error a possibility for the finding in this study?

Since P is greater than 0.05, the reduction in headache frequency with the new drug compared to placebo is *not* statistically significant. This means that the null hypothesis of no difference between study treatments is accepted.

Yes, type II error is a possibility for nonstatistically significant findings. Note that a reduction of 40% in headache frequency appears to be a fairly substantial treatment effect, even though it was not statistically significant. Concluding there was no difference between the new drug and placebo might be a wrong conclusion here (i.e., a type II error). Power provides the likelihood of not making a type II error (β). This study did not report power, so we do not know if it was acceptable (80% or higher). Thus, the actual likelihood of type II error is unknown. Since there was a rather small number of study patients (small sample size), the power could be low. A larger study should be performed.

Statistical significance versus clinical significance (importance)

There is a difference between statistical significance and clinical significance. As indicated previously, statistical significance is present when the P value is less than a designated cut-off, usually 0.05. This means that the likelihood that chance alone was responsible for the effect seen is acceptably low, and the null hypothesis is rejected. But, just because a difference between treatments in a study is not likely a chance finding, does that automatically mean that the difference is large enough to be clinically important? Not necessarily. A study with a very large number of patients may find a fairly small treatment difference to be "statistically significant" (unlikely to have resulted from chance). Let's say a large study finds a difference in diastolic blood pressure between two treatments of 2.5 mmHg to be statistically significant. Would a difference of 2.5 mmHg in diastolic blood pressure be of clinical importance for most patients? Probably not. When $P < 0.05$ and the finding is statistically significant, *always* look at the actual treatment effect or difference to determine clinical significance.

Key Points

How should we interpret a "statistically significant" finding?

- "Statistical significance" simply means that a study's finding was not likely to result from chance; the null hypothesis of no treatment difference is rejected.
- A finding is statistically significant when $P < 0.05$ (sometimes a smaller P value might be used, but a cut-off of 0.05 is most common).

Can a study finding be concluded to be clinically significant if it is *not* statistically significant?

No. Remember the following:

- A study finding can be statistically significant but the actual effect might be too small to be of clinical importance in practice.

- In order to conclude that a study finding is clinically important there needs to be statistical significance. Without statistical significance, the finding has a higher than acceptable likelihood that it resulted from chance alone and not from a treatment effect. Thus, no clinical conclusions can be made in the absence of statistical significance.
- If a study finds a difference large enough to be of clinical importance, but it is not statistically significant, consider whether a type II error occurred – meaning, a real treatment effect was *missed*. Most likely this type of error would result from insufficient power in a study and too small a sample size. When a study finds a large treatment difference that is not statistically significant, especially when power is not reported or is too low, further study is needed.

Can a study finding be statistically significant but *not* clinically significant?

Yes. With a large number of patients, a small treatment effect might be statistically significant (not likely to result from chance) but too small to be of clinical importance in actual practice.

How can one determine if a reported CI is statistically significant?

If a CI includes the value that indicates no treatment effect, the CI will not be statistically significant. For a difference in a specific outcome measure (e.g., blood pressure following therapy, measured lab values before and after therapy), the value of 0 would equal no treatment effect. For example, suppose a study reported cure rates from drug A and drug B of 72% and 68%, respectively, with a difference between cure rates of 4% (95% CI = −2% to 10%). Note that the value of 0 (no difference in cure rate between drug A and B) is within this CI range; thus, the actual population difference could be 0. This CI range would not be of statistical significance (one does not even need to look at the *P* value to tell this).

Measures of risk, risk reduction, and clinical utility

Clinical studies often look at the likelihood of an event occurring with therapy, either an adverse effect or a beneficial outcome such as prevention of a stroke, blood clot, or relapse of a medical condition. When comparing two therapies (treatment versus placebo, treatment versus active control), clinicians would like to know the extent to which the treatment might reduce the likelihood of an adverse event, so they can determine how best to apply the results to patients. There are several measures that can be used to accomplish this, referred to here as measures of risk, risk reduction, and clinical utility. These measures are: *odds ratio* (OR), *relative risk* (RR), *relative risk reduction* (RRR), *absolute risk reduction* (ARR), *number needed to treat* (NNT), and *number needed to harm* (NNH).

Most of us are familiar with the term "odds," particularly related to gambling or games such as the lottery. The *odds* of an event occurring is calculated as the number of times the event occurs divided by the number of times it does not occur. For example, the odds of rolling the number six with one throw of a die is the number of times a six would occur on the die/number of times a six would not occur on the die, which in this case would be 1 time a six would occur/5 times a six would not occur = 1/5 or 1 to 5. The OR, which is often reported in clinical studies, is a *ratio of the odds of an event occurring in one treatment group divided by the odds of the event occurring in the other group*. Usually the *numerator* of the OR is the odds in the treatment group (or in the cases for a case-control study) and the *denominator* is the odds in the control group (note: sometimes a study might reverse these, so pay attention to what is in the numerator and denominator since it will affect the interpretation). Thus, the OR is interpreted as follows:

- OR < 1 (the odds of the event occurring in the treatment group is less than the odds in the control group, assuming treatment is in the numerator).
- OR = 1 (the odds of the event occurring in the treatment group is equal to the odds in the control group – both the numerator and denominator are equal).
- OR > 1 (the odds of the event occurring in the treatment group is greater than the odds in the control group, assuming treatment is in the numerator).

For example, suppose a study examined the efficacy of a new drug for migraine prophylaxis compared with propranolol (control). A total of 350 patients were randomly assigned to receive either the new drug (150 patients) or propranolol (200 patients). The patients were followed over 2 weeks and the outcome measure was migraine development. A total of 30 patients (20%) in the new drug group and 50 patients (25%) in the propranolol group had a migraine over the 2-week study period. What is the OR for migraine development with the new drug compared to propranolol?

$$\frac{\text{odds of migraine with new drug}}{\text{odds of migraine with propranolol}}$$
$$= \frac{30 \text{ (\# with migraine)}/120 \text{ (\# with no migraine)}}{50 \text{(\# with migraine)}/150 \text{(\# with no migraine)}} = \frac{0.25}{0.33} = 0.76$$

Since the OR is <1, the odds of a migraine developing with new drug therapy are less than with propranolol (the odds with new drug therapy are actually 76% of the odds with propranolol).

Now let's talk about *risk*. The *risk* of an event occurring is calculated as *the number of times the event occurs divided by the total number of persons in the involved (exposed) group*. Note that the difference between calculating risk or odds is in the denominator. Using the previous die example, the risk of rolling the number six with one throw of a die is the number of times a six will occur/total number of numbers present on die, which in this case would be 1 time a six would occur/total of six numbers on the die = 1/6 or 0.17. The *RR*, which is often reported in clinical studies, is a *ratio of the risk of an event occurring in one treatment group divided by the risk of the event occurring in the other group*. Usually the *numerator* of the RR is the risk in the treatment group and the *denominator* is the risk in the control group (sometimes a study might reverse these, so pay attention to what is in the numerator and denominator since it will affect the interpretation). Thus, the RR is interpreted as follows:

- RR < 1 (the risk of the event occurring in the treatment group is less than the risk in the control group, assuming treatment is in the numerator).
- RR = 1 (the risk of the event occurring in the treatment group is equal to the risk in the control group – both the numerator and denominator are equal).
- RR > 1 (the risk of the event occurring in the treatment group is greater than the risk in the control group, assuming treatment is in the numerator).

Let's use the same study example shown earlier: 350 patients were randomly assigned to receive either the new drug (150 patients) or propranolol (200 patients). The patients were followed over 2 weeks and the outcome measure was migraine development. A total of 30 patients (20%) in the new drug group and 50 patients (25%) in the propranolol group had a migraine over the 2-week study period. What is the RR for migraine development with the new drug compared to propranolol?

$$\frac{\text{risk of migraine with new drug}}{\text{risk of migraine with propranolol}}$$

$$= \frac{30 \text{ (\# with migraine)}/150 \text{ (\# in new drug group)}}{50 \text{(\# with migraine)}/200 \text{(\# in propranolol group)}} = \frac{0.20}{0.25} = 0.8$$

Since the RR is < 1, the risk of a migraine developing with new drug therapy is less than with propranolol (the risk with new drug therapy is actually 80% of the risk with propranolol).

Key Point

What is the difference between the OR and RR? Are they both saying the same thing?

Although they are both measures for comparing the likelihood of an event occurring (or not occurring) between two groups, they are *not* saying the same thing:

- The OR gives the relative odds of the event occurring, and the RR is looking at the probability (risk) of an event occurring.
- There can be large differences between both measures depending on the extent to which the event has occurred (see below).
- For case-control study designs, only the OR makes sense. Why? Think back to the design of a case-control study (see Chapter 5). With a case-control study, the cases are patients who already have the event or outcome and the controls are another group of similar patients who lack the event or outcome. To calculate risk, one needs to know the *total number exposed* to the therapy (i.e., treated) for the denominator. That is unknown for a case-control design.

- Either an OR or RR can be used for designs other than the case-control, but the results need to be interpreted correctly.

For example, suppose that for every 100 patients who go to their doctor with a sudden onset of blurred vision, 60 had a stroke. The "risk" that the blurred-vision patients had a stroke is 60/100 = 0.6 or 60%. But, the "odds" that these persons had a stroke is 60/40 = 1.5, a fairly big difference from 0.6. The 1.5 is saying the odds of the blurred-vision patients having had a stroke is 1.5 times the odds that they did not.

Now let's make the stroke rates much smaller. Now, for every 100 patients who go to their doctor with a sudden onset of blurred vision, 20 had a stroke. The "risk" that the blurred-vision patients had a stroke is 20/100 = 0.2 or 20%. The "odds" of these persons having had a stroke is 20/80 = 0.25, much closer in value to the risk of 0.2.

Bottom line

The more frequently the event occurs (i.e., the larger the actual occurrence rate), the larger the difference will be between a calculated OR and RR. There is one study design (case-control) that needs to report the OR. For the other study designs, understand how to interpret the result for the OR or RR.

Key Point

RRR can be deceiving because, since it is a proportion (percentage), it can have the same value for large or small actual event rates. Consider the following two examples:

- Suppose ulcers occur in 5.3% of patients taking drug A and in 2.3% of patients taking drug B. What is the RRR for ulcer development with drug B compared to drug A? RRR = (5.3% − 2.3%)/5.3% = 0.57 = 57% (RRR is usually expressed as a percentage).

- Suppose ulcers occur in 53% of patients taking drug A and in 23% of patients taking drug B. What is the RRR for ulcer development with drug B compared to drug A? RRR = (53% − 23%)/53% = 0.57 = 57%.

- The RRR is identical in both instances, even though the actual difference between the ulcer incidence rates is only 3% for the first example but 30% in the second example. This illustrates *the problem with the RRR: it does not differentiate between small and large actual event rates.*

This leads into another commonly used measure of risk, the RRR. This is simply the extent of the reduction in the relative risk. An analogous way to think of this is as follows: suppose you go shopping and note that the regular price of a shirt/blouse you would like to buy is $25 but the sale price is $18. You just love a good sale. How much percentage reduction are you getting with this sale? In other words, relative to $25, how much smaller (proportionally, in percentage) is the $18 price? This "RRR" can be calculated in two ways: (1) 1 − RR; or (2) (control rate − treatment rate)/control rate. In the migraine example, the RRR for migraine development with the new drug compared to propranolol is 1 − 0.8 (RR) = 0.2 or 20%. For the shirt/blouse example, the second equation might be simpler and would be ($25 − $18)/$25 = 0.28 or a 28% reduction (sale). Similar to why shoppers like to see the percentage sale reduction they are receiving for a purchase, the RRR has value for clinicians because it provides an indication of the proportionate reduction in risk that patients might experience. For example, if a patient is at high risk for experiencing a serious diabetes complication and a drug could reduce the risk of developing this complication by 75%, that could be very clinically important for the patient.

Key Point

Sometimes a study will provide a "risk reduction" without indicating whether it is an RRR or an ARR. Is there a reason why investigators might want to do this? Yes, there can be. Suppose one or more of the investigators are employed by the drug's manufacturer. They might want to present their findings in a manner that looks the "best" to readers. Their study drug might cause hypoglycemia in 1.5% of patients compared to 3% with placebo, and the investigators report a hypoglycemia risk reduction of 50% for the study drug. While this appears very impressive at first glance, the 50% is an RRR with only a 1.5% ARR. If the hypoglycemia was mild, the small actual reduction would be of minimal clinical importance. Thus, *always check to see whether a risk reduction reported in a study is a RRR or an ARR.*

Is there a measure that provides the actual difference in the event rates between treatments? Yes, the ARR gives the actual difference in rates. Simply subtract the percentage event rate with one treatment from the percentage event rate in the other treatment group. For the migraine example in which a migraine occurred in 20% of patients taking the new drug and 25% of patients taking propranolol, the ARR = 25% − 20% = 5%. It is always good to check the ARR (easy to calculate) to see the actual treatment effects, particularly when an RRR (proportional reduction) is reported.

Obviously not all patients are going to benefit from any given therapy to the same extent. In the migraine example given earlier with the new drug and propranolol, a migraine occurred in 20% of the new drug group compared to 25% of the propranolol patients during the study period. Since each of these therapies might have its own unique adverse effects, clinicians also need to weigh the benefits versus risks of one therapy against the other. Thus, it would be helpful to consider how many patients would need to be treated with the new drug, instead of propranolol, to prevent one additional migraine headache over the 2-week period. The NNT tells us that. It is easily calculated as NNT = 1/ARR. In the migraine example, the ARR with the new drug compared to propranolol is 25% (propranolol migraine %) − 20% (new drug migraine %) = 5% = 0.05 (*always* make sure to use the decimal instead of the percentage value − otherwise, you will end up with an NNT of <1 or a partial patient!). Thus, the NNT = 1/0.05 = 20. This means that 20 patients would need to be treated with the new drug instead of propranolol to prevent one additional migraine episode. The NNT allows a clinician to consider the possible risks patients would be exposed to from a treatment before seeing the benefit.

Key Points

- *If the NNT is not a whole number, always round up any decimals to the nearest whole number.* For example, suppose the ARR for ulcer development with drug A (4.3% ulcer incidence) compared to drug B (6.9% ulcer incidence) is 2.6% (6.9% − 4.3% = 2.6%), and the calculated NNT = 1/.026 = 38.46. The NNT here would be 39 patients (rounded *up*).
- *The smaller the difference in event rates between treatment groups, the larger the NNT will be.* Suppose two different studies comparing drug A and drug B reported the following ARRs: study 1 = 6.7% ARR and study 2 = 14.8% ARR. The NNT in study 1 would be 1/0.067 = 14.93 = 15 and the NNT in study 2 would be 1/0.148 = 6.76 = 7.
- *The smaller the NNT, the smaller the number of patients that will need to be treated (over the duration of time used in the study) to prevent one additional adverse event.* Let's take the previous example, but suppose this time the ARRs are as follows: study 1 = 6.7% and study 2 = 0.14%. The NNT in study 1 is 15 but the NNT in study 2 is now 1/0.0014 = 714.29 = 715. This makes intuitive sense, in that the smaller the treatment difference, the more patients would need to be treated with one drug instead of another to demonstrate that potential advantage.
- *There is no "cut-off" for how small the NNT should be before it is considered important or "significant."* One must always consider the benefits and risks of the treatments to determine whether a certain NNT justifies treatment. For example, if NNT = 500 but the outcome (that would be prevented) is very serious and any adverse effects from therapy are very mild, therapy might still be justified. On the other hand, if the NNT = 20 but the adverse outcome to be prevented is very minor and potential adverse effects from therapy are very serious, therapy might not be justified.

Worked example

Example 9.9

A study compared the incidence of severe hypoglycemia in diabetic patients who received either an oral drug to treat diabetes (238 patients) or a new inhaled insulin product (295 patients). At the end of a year, 65 patients (27.3%) who received the oral drug and 103 patients (34.9%) who received the inhaled insulin developed at least one severe episode of hypoglycemia.

What is the OR for development of severe hypoglycemia with the oral drug compared to inhaled insulin?

The OR = odds of severe hypoglycemia with the oral drug/odds of severe hypoglycemia with inhaled insulin = 65 (number of patients with hypoglycemia with oral drug)/173 (number of patients without hypoglycemia with oral drug)/103 (number of patients with hypoglycemia with inhaled insulin)/192 (number of patients without hypoglycemia in inhaled insulin group) = 0.376/0.536 = 0.70. Since 0.7 is less than 1, it means that the odds of severe hypoglycemia with the oral drug is 0.7 times the odds with inhaled insulin.

What is the RR for development of severe hypoglycemia with the oral drug compared to inhaled insulin?

The RR = risk of severe hypoglycemia with the oral drug/risk of severe hypoglycemia with inhaled insulin = 65 (number of patients with hypoglycemia with oral drug)/238 (number of patients in oral drug group)/103 (number of patients with hypoglycemia with inhaled insulin)/295 (number of patients in inhaled insulin group) = 0.273/0.349 = 0.78. Since 0.78 is less than 1, it means that the risk of severe hypoglycemia with the oral drug is 0.78 times the risk with inhaled insulin.

What is the NNT for the development of severe hypoglycemia with the oral drug compared to inhaled insulin? Interpret precisely what this NNT value means.

ARR = 34.9% − 27.3% = 7.6%. NNT = 1/ARR = 1/0.076 = 13.16 = 14. This NNT means that 14 patients need to be treated with oral insulin instead of inhaled insulin for a year (the duration of the study) in order to prevent one case of severe hypoglycemia.

Key Points

- Any time more adverse events occur with a treatment compared to the control, an ARI would be used to describe the actual increase in event rates with the treatment. To keep from being confused, just think about the actual study findings. It makes no logical sense to report an ARR for a therapy that has more adverse events than the control group.
- When an ARI is present, calculate an NNH instead of the NNT.

Now suppose that investigators are studying a new therapy and find that it is worse, not better, than the control. Let's take the migraine example from earlier, but this time the percentages are reversed: A total of 350 patients were randomly assigned to receive either propranolol (150 patients) or the new drug (200 patients). The patients were followed over 2 weeks and the outcome measure was migraine development. A total of 30 patients (20%) in the propranolol group and 50 patients (25%) in the new drug group had a migraine over the 2-week study period. Obviously, in this case, there is not an absolute risk *reduction* with the new drug compared to propranolol. Rather, the risk of migraine is *increased* with the new drug. In this case, we use a term called *absolute risk increase* (ARI) to describe the risk of migraine for the new drug compared to control. It is determined in a similar way as ARR, by simply taking the migraine percentage in the new drug group and subtracting the percentage in the propranolol (control) group. In this example, the ARI = 25% − 20% = 5%. Instead of calculating the NNT, we now calculate an analogous NNH. The NNH is defined as the number that need to be treated with one therapy instead of the other in order to *cause* one additional adverse outcome. It is calculated as 1/ARI. Thus, in our example here the NNH is 1/0.05 = 20. This means that 20 patients would need to be treated with the new drug instead of propranolol in order to cause one additional migraine episode.

Drop-outs and data handling

Chapter 7 described the use of outcome measurements to determine treatment effects. In an ideal study, every patient who is enrolled would complete the study and the investigators would be able to obtain all the necessary data from every patient. Unfortunately, there usually is no such thing as an "ideal study." Many times patients will quit a study (drop out) for a variety of reasons, some of which might have nothing to do with the study itself. For example, patients might move out of the area or simply grow tired of showing up for all the necessary follow-up visits. In other cases, patients might develop adverse effects and decide to quit the study. The question arises, how should a study's data be analyzed when not all of the patients complete all parts of the data collection?

Key Point

With an intent-to-treat analysis, how does a study include data from patients who dropped out and are no longer available for subsequent outcome measurements? There are different approaches that studies can use to include measures from patients no longer in a study. These often include the following:

- Last observation carried forward – the last measure the study obtained from a patient is used for subsequent measurements after they drop out.
- Some studies that have as an outcome measure "success" or "failure" will assume that a patient has "failed" therapy after dropping out.
- Imputation of data – missing data points are filled in by using various procedures or equations that estimate what the values would be if the patients remained in the study.

There are two commonly used methods for handling the data from drop-outs in a study: (1) *intent-to-treat* (*intention-to-treat*), and (2) *exclusion of subjects* (*per protocol*). With *intent-to-treat*, the data from all patients randomized into a treatment group are analyzed in the study regardless of whether they completed the study. A proposed advantage of using this data-handling method is that it might better represent clinical practice in that some patients in real life will quit taking their therapy or not adhere to the therapy as directed. Thus, if a study includes these types of patients, the efficacy observed would mimic actual practice.

With the *exclusion of subjects* method, only those patients who complete the study protocol as specified would be included in the data analyses. This method excludes those patients who dropped out before the end of the study for any reason or those who deviated from the protocol, e.g., skipping follow-up measurements.

A proposed advantage of the exclusion of subjects data-handling method compared to intent-to-treat is that exclusion of subjects can better determine the actual efficacy of the therapy, especially if a significant number of patients drop out due to reasons unrelated to the therapy, e.g., lost to follow-up. Why? Suppose that 100 patients are enrolled in a study and receive drug A. A total of 35 patients dropped out during the study for a variety of reasons. Overall, it was found that 45 of the 100 patients had a favorable response to drug A. What would be the efficacy if the intent-to-treat data-handling method were used? The efficacy would be 45/100 (data from all patients are included regardless of whether or not they dropped out) or 45%. What would be the efficacy if exclusion of subjects were used for the data-handling method? The efficacy would be 45/65 (data from only those patients who completed the study were included in the analysis) or 69%. Note that drug A efficacy is higher with the exclusion of subjects method because the denominator is smaller when the drop-outs are removed.

Sometimes studies will refer to a "*modified intent-to-treat*" data-handling method. Thus usually means that patients needed to have taken one dose of the therapy or completed one data collection. Although fairly frequently used, a modified intent-to-treat method is not the same as the intent-to-treat method. For more information about intent-to-treat analyses, the reader is referred to a review by Moncur and Larmer (2009).

Key Points

- *The intent-to-treat data-handling method is preferred over exclusion of subjects.* With exclusion of subjects, it is possible that bias might be introduced by who drops out of the study, e.g., patients with certain characteristics might quit the study to a greater extent than patients who lack these characteristics. The benefits of random assignment might be reduced if patients drop out of the study in a nonrandom manner.

- *Studies will often use both the intention-to-treat and exclusion of subjects data-handling methods to analyze study findings when they have a considerable number of drop-outs.* This is desirable because it allows readers to determine the effect that each data-handling method has on the therapy efficacy rates, and allows the advantages of each method to be considered.

- *Determine the resulting effect on sample size, and therefore statistical power, when the exclusion of subjects data-handling method is used.* Keep in mind that when there are a large number of study drop-outs and these patients are removed from the study analyses, this will reduce the sample size and the resulting statistical power.

Worked example

Example 9.10

A 1-year study compared Fosamax (200 patients) with Actonel (200 patients) for preventing bone fractures in patients with osteoporosis. It was determined that the power of the study to detect a 5% difference in the incidence of bone fractures, with 400 total patients and an $\alpha = 0.05$, was 80%. Sixty patients (30%) dropped out of the Fosamax group and 10 patients (5%) dropped out of the Actonel group. The number of patients with new bone fractures was determined in both treatment groups.

What was the effect size used for the power analysis?

The effect size, the difference between treatments that the investigators would like to be statistically significant if present, is a 5% difference in the incidence of bone fractures.

Which data-handling method, intention-to-treat or exclusion of subjects, would reduce the statistical power for comparing the bone fracture rates between treatment groups?

The exclusion of subjects method would reduce the statistical power since a total of 70 patients dropped out of the study, and these patients would not be included in the bone fracture analysis.

Discussion section

Busy clinicians will often skim a published study to learn about its key features and findings; if of interest, they will then read the study in more detail. The part of the publication that clinicians will usually read is the discussion section. This section should ideally summarize all the important findings of the study and analyze the results in relation to what previous studies or other relevant literature on the subject had reported.

Unfortunately, the study's authors can sometimes use this section to focus on only their positive results while glossing over possible therapy adverse effects. They can neglect to mention important limitations of their work that readers should be aware of. The authors might also inappropriately extrapolate their findings to situations outside the objectives or scope of the study, or to a population that was not defined by the study's eligibility criteria. Chapter 6 discussed the potential conflicts of interest for a study's investigators. The discussion section of a published study is one of the areas in which bias can be introduced by the investigators.

The journal *Annals of Internal Medicine* in its "Information for Authors" provides recommendations for the types of information the discussion section should include (accessed at: http://annals.org/public/authorsinfo.aspx #manuscript-preparation, July 2012). Readers should look for these points when analyzing the discussion section of a published study:

- a brief synopsis of key study findings, with analysis of how those findings contribute to or expand upon what is already known on the subject
- possible reasons or explanations for the results found
- discussion of how the study findings support or differ from the results seen with other relevant published trials. Tables and figures should be used when possible to allow readers to compare key aspects of the available relevant studies easily
- an honest analysis of any limitations of the study and how the study attempted to avoid or minimize those problems
- the types of future research needed in the subject area
- a final summary that clearly states the study's conclusions and the potential generalizability and clinical applicability of the study findings.

A summary of key points and how to apply the information from this chapter to practice follow.

Key Points

- The mean and median are the two most commonly used measures to indicate a "typical" patient's response to therapy. The mean is often used for ordinal-level data, although it is difficult to determine exactly what the decimal means for nonwhole numbers – use caution in this instance.

- Any time a mean is reported, look for the SD to provide an indication of the spread of the individual patients' responses for that measure. A small SD indicates that most of the individual responses were tightly clustered around the mean value.
- Any time a median is reported, look for an IQR to provide an indication of the spread of the individual patients' responses for that measure. A small IQR indicates that 50% of the individual responses were within that narrow range of values between the 25th and 75th percentiles.
- SEM should not be used as a measure of variability in clinical studies; SD should be provided instead. SEM is *always* smaller than SD, so investigators will sometimes report SEM to minimize the apparent variability of the individual patient responses in their study.
- The CI provides a range of values likely to contain the actual population value for that measure, at the level of confidence specified.
- Even with a 95% CI, there is a 5% likelihood that the range of values reported will not contain the actual population value. But this is an acceptable level of confidence.
- The CI width is affected by the level of confidence selected (90% versus 95% versus 99%), the sample size, and the underlying SD for that measure.
- The P value provides the likelihood (probability) that the study results seen would have occurred if the null hypothesis were true; it gives the probability that the difference observed in the study was simply due to chance and not from the treatment(s).
- Alpha (level of significance) is the likelihood of making a type I error and is usually set at 0.05 for power and statistical calculations; a P value of less than 0.05 is considered statistically significant.
- Beta is the likelihood of making a type II error and is acceptable when less than 0.2 (20%).
- Power is defined as the likelihood of *not* making a type II error; it is calculated as $1 - \text{beta}$ and is acceptable when $\geq 80\%$ (0.8).
- The easiest factor that investigators can change to increase a study's power is the sample size. Increasing the sample size will increase power. Anything that decreases the sample size during the study, e.g., comparing smaller subgroups of patients within the study, patient drop-outs that are not included in the data analyses, will decrease the power calculated at the start of the study.
- A finding *cannot* be concluded to be clinically significant unless it is first statistically significant. However, a finding can be statistically significant but too small to be of clinical importance.
- The OR, RR, RRR, ARR, and NNT are measures used to describe the risk, risk reduction, and clinical usefulness of therapy.
- The OR gives the ratio of the odds of an event occurring, and the RR gives the ratio of the risk of an event occurring. Both measures have the same numerator: the number of persons in the treatment group that have an event. They differ in the denominator: the denominator for the OR is the number of persons in the treatment group that do *not* have the event, while the denominator for the RR is the total number of persons in the treatment group.
- Only the OR should be reported for a case-control study.
- The ARR provides the actual difference (reduction) in event occurrence between treatment groups. The RRR provides the relative (proportional) reduction in event occurrence between treatment groups.

- The NNT is calculated as 1/ARR. It should *always* be rounded up to the nearest whole number.
- If a treatment has a greater incidence of an adverse event than the control group, an ARI would be reported for that treatment. An NNH can be determined from the ARI, calculated as 1/ARI.
- Always look at the data-handling method used in a study; the intent-to-treat method is preferred to minimize the possibility of bias and to mimic actual practice.
- If there are a significant number of drop-outs in the study, the study should best analyze their findings using both the intent-to-treat and exclusion of subjects methods.
- The discussion section of a published study should summarize its important results. The results should be discussed in light of the findings from other relevant studies reported in the literature, including similarities and any important or unexpected differences. Both the strengths and weaknesses of a study should be addressed. How the results could be applied to clinical practice or extrapolated to patients outside the study should also be included.

How to apply to practice

- Use the median instead of the mean when interpreting the central tendency (clustering) for skewed data (e.g., when both the median and mean are given and appear substantially different).
- Look at the value of a SD or IQR provided; a wide spread indicates that the study patients had a substantial variability in their responses to that outcome measure. Thus, it might be hard to predict how patients outside the study would respond to therapy.
- Always convert any SEM values reported in a study to SD, using the equation SEM = SD/$\sqrt{n}$; then use the SD to examine the variability (spread) of individual patient responses around the mean value.
- When SEM is given in a study, consider whether there are any potential conflicts of interest for the investigators that might explain why they chose to use that measure instead of the SD.
- Always use the CI to indicate what the population value is likely to be. If all the values within the CI range are likely to be clinically important in practice, then you have a good indication that the finding will be of clinical significance in the population.
- If a CI includes the value that indicates no treatment effect (0 for a treatment difference or change in outcome measure; 1 for a RR or OR), the CI is *not* statistically significant.
- The smaller the *P* value, the less likely it is that chance alone was responsible for the finding.
- If $P < 0.05$ (unless the investigators indicate otherwise), the finding is statistically significant and the null hypothesis is rejected.
- If $P \geq 0.05$, the finding is not statistically significant (the null hypothesis is accepted) and consider whether a type II error might have occurred.
- Look for a reported power of at least 80% in a study. If the power is not adequate, it is most likely because not enough patients were included in the study.
- For statistically significant findings, look at the actual size of the treatment effect. If large enough to be of clinical importance, the finding is clinically significant.
- If a nonstatistically significant finding looks large enough to be of clinical importance, consider whether a type II error occurred (missing a real treatment

effect). Make sure the power for that outcome measure was at least 80%. If the power was <80% or not reported (either missing or calculated for a different outcome measure), type II error is a potential problem.

- When a study reports an OR or RR, always look to see whether the treatment or control data are in the numerator or denominator. When the treatment group is in the numerator and the control group is in the denominator (the usual situation), the odds or risk of the event occurring is less in the treatment group when the value is <1, and greater in the treatment group when the value is >1. If the control group is in the numerator and the treatment group is in the denominator, the odds or risk of the event occurring is less in the control group when the value is <1, and greater in the control group when the value is >1.

- If a study reports only a "risk reduction," determine whether it was a RRR or an ARR. If a RRR is reported, always look at the ARR to see if the actual treatment difference is of possible clinical importance.

- Look for, or calculate, the NNT (= 1/ARR) to determine the possible risks patients would be exposed to from a treatment before seeing the benefit. The larger the NNT, the larger the number of patients that would be exposed to possible adverse effects without necessarily receiving a treatment benefit.

- If the intent-to-treat data-handling method is used to analyze the results in a study with a fairly large number of patient drop-outs, the efficacy of the drug might appear lower than it actually is.

- If the exclusion of subjects data-handling method is used to analyze data with a fairly large number of patient drop-outs, fewer patients will be included in the analyses so the power could be adversely affected. Make sure the power would still be acceptable.

- Carefully read the discussion section of a published study to determine if biased statements were made, adverse effects were minimized, or inappropriate extrapolations were stated. This might be more likely when there are potential conflicts of interest for the study's investigators.

Self-assessment questions

Question 1
Which of the following measures is also described as the "50th percentile"?

a. Mean
b. Median
c. Mode

Question 2
A study reports that the median blood concentration in persons who overdosed on drug X was 230 mg/L, with a mean concentration of 398 mg/L. What conclusion should be drawn about the data given the median and mean concentrations reported? What measure of variability should be reported in this circumstance?

Question 3
Study A reports that the median estrogen concentration in 100 women using an oral contraceptive (OC) is 256 ng/mL and the mean concentration is 380 ng/mL. Study B

similarly reports that the mean estrogen concentration in 150 women using the same OC is 380 ng/mL but does not report the median. Which of the following can be correctly concluded from these data?

a. If a few extreme outlying concentrations existed in study A, only the median was unaffected by them.
b. It was not appropriate for study A to report a median value for continuous-level data.
c. The individual patients in study A had estrogen concentrations that were similar to those in the patients in study B.
d. The individual patients in study A had estrogen concentrations that were very different than those in the patients in study B.

Question 4

A study looked at the effects of the addition of an angiotensin-converting enzyme inhibitor (ACEI) to a low-sodium diet on proteinuria in diabetic patients. After 8 weeks of therapy, it was found that the addition of the ACEI reduced the extent of proteinuria by 49% (**95% CI = 32–66%**). Interpret the meaning of the statement in bold (include the numbers provided in the statement in your answer).

Question 5

A study compared the efficacy of lisinopril (L), enalapril (E), and valsartan (V) for preventing renal damage in diabetic patients. A total of 548 patients were randomized to receive L (184 patients), E (182 patients), or V (182 patients) for 6 months; the primary outcome measure was change (increase) in serum creatinine (SCr) concentrations. To detect a difference between groups of 10% in the number of patients with an increase in SCr, 175 patients were needed per group with an $\alpha = 0.05$ and $\beta = 0.19$. During the study, several patients dropped out for various reasons. After 6 months, the following results were obtained for the number of patients with an increase in SCr:

Group L: 29 of 160 patients (18%); group E: 16 of 155 patients (10%); group V: 35 of 165 patients (21%).

Differences between groups were: L versus E (8%; $P = 0.06$); L versus V (3%; $P = 0.23$); E versus V (11%; $P = 0.042$).

Side-effects occurred in: 15% of L patients, 21% of E patients, and 7% of V patients.

Differences in side-effects between groups were: L versus E (6%; $P = 0.04$); L versus V (8%; $P = 0.036$); E versus V (14%; $P = 0.021$).

a. List those P values that were statistically significant.
b. What was the calculated power in this study? Is the power acceptable?
c. Was type II error a possibility for the comparison of L versus E with regard to the number (%) of patients with an increase in SCr? Explain.
d. Suppose the investigators wanted to increase the power of their study. What would be the best way to accomplish this?

Question 6

The following two studies were published:

Study A: 100 patients received a new cholesterol-lowering drug, and a cholesterol concentration during therapy of 185 mg% ± 5 mg% (mean ± SEM) was reported.

Study B: 100 patients received this same new drug, and a cholesterol concentration during therapy of 185 mg% ± 18 mg% (mean ± SD) was reported.

a. Which study had the *greatest* individual patient variability in serum cholesterol concentrations, study A or study B?
b. Explain *precisely* and *completely* what the ± **18 mg%** refers to in study B.

Question 7

Toradol was compared with naproxen for treating acute low-back pain. Complete pain relief was experienced in 58% of patients receiving toradol and 46% of patients taking naproxen (difference = 12%, **95% CI = −7% to 31%**). The degree of back tenderness and the extent of flexibility were also measured. Which of the following is correct concerning the bolded term?

a. 95% confidence that the difference in pain relief between toradol and naproxen in the study patients fell between −7% and 31%
b. 95% confidence that the population difference in pain relief between toradol and naproxen might range from 7% fewer patients to up to 31% more patients experiencing pain relief with toradol
c. 95% of the toradol patients and naproxen patients experienced pain relief from 7% to 31% of the time

Question 8

Would the CI reported in question 7 be statistically significant?

Question 9

A double-blind randomized study compared losartan (347 patients) with atenolol (320 patients) for the treatment of hypertension in diabetic patients. During the 12-month study, 24 (6.9%) of the losartan-treated patients developed kidney disease compared to 40 (12.5%) of the atenolol-treated patients.

a. What is the OR for kidney disease development in patients treated with losartan compared to atenolol?
b. What is the RRR for kidney disease development in patients treated with losartan compared to atenolol?
c. What is the NNT for kidney disease development in patients treated with losartan compared to atenolol? Describe *exactly* how this NNT value should be interpreted.
d. What is the RRR for kidney disease development in patients treated with losartan compared to atenolol?

Question 10

A 24-week study compared placebo (600 patients) with drug X (598 patients) for preventing retinopathy in diabetic patients. A total of 206 patients (34%) taking drug X and 175 patients (29%) taking placebo dropped out of the study for a variety of reasons, most not related to the therapy. Data were analyzed for all 1198 patients who were enrolled in the study. At the end of the study, a total of 9% of placebo patients compared to 7% of drug X patients developed retinopathy. Which data-handling method was used in this study? How would this data-handling method likely have affected the drug X efficacy reported?

Summary

This chapter covered the key points needed to review and interpret the results reported in a clinical study, as well as the important aspects that should be included in the discussion section of a published study.

10

Evaluating clinical studies – step 6: putting it all together

Learning objectives

Upon completion of this chapter, you should be able to:
- identify the important questions to ask when analyzing and critiquing a published clinical study
- summarize the strengths and weaknesses of a published clinical study and provide an appropriate conclusion that considers any key limitations identified, the role of the therapy in practice, and whether further research is required.

Introduction

Previous chapters reviewed many important factors to consider when reading a published study and determining how best to apply the results in practice. Important considerations included the: strengths and limitations of various study designs, e.g., case-control and cohort, as well as the types of controlled experimental designs; journal and authors involved; introduction/background and study rationale provided; identification and enrollment of the study sample patients/subjects; treatment regimens used (control and experimental); outcome measures used; data handling and statistical methods employed; presentation and interpretation of the results; and authors' discussion and conclusion. This chapter provides questions and considerations that should be addressed when critiquing published clinical studies, based on the concepts covered previously. The focus of this chapter is on controlled experimental studies given their importance in establishing and comparing therapy efficacy. Refer to Chapter 5 of this book for the advantages and disadvantages of the other types of study designs, including case-control, cohort (prospective and retrospective), cross-sectional, and n-of-1 studies.

Key questions and considerations when critiquing published experimental studies

(Note: If the answer is "no" to any of the items in bold type, that weakness alone might be enough to invalidate the study – extra caution must be used before applying results from that study to practice.)

Journal and authors/investigators (refer to Chapter 6)

- Does the journal have an editorial board consisting of individuals with expertise in the involved subject area? Does the journal use peer review for the papers published?
- Are there apparent potential conflicts of interest for the authors/investigators? If potential conflicts of interest exist, did they appear **actually to affect** the study's objective, methods, or conclusions?

Introduction/background (refer to Chapter 6)

- Was an appropriate scientific background and rationale provided?
- Is the stated objective or hypothesis consistent with the research question that needed to be addressed?
- **Is the study adequately designed to fulfill its stated objective or hypothesis?**

Patients/subjects (refer to Chapter 7)

- Were the inclusion and exclusion criteria appropriate and representative of the population of interest? Were factors (e.g., nonstudy drugs, concurrent medical conditions) that might inappropriately interfere with the study excluded (to the extent practical and feasible)? If not, how does this limit the population to which the study's findings could be applied?
- Does the way in which the patients were enrolled (i.e., sampled/selected) limit the population to which the findings could be applied?
- Was the number of patients enrolled and analyzed sufficient to maintain at least 80% power for the outcome measurements?

Treatment regimens (refer to Chapter 7)

- Was the control used appropriate for the study's purpose (e.g., placebo for determining therapy efficacy, active control for determining comparative efficacy)?
- Were the dosing and administration regimens appropriate and representative of what would be used in clinical practice?
- Was a concurrent control design used (preferred)? If not, were sufficiently long wash-out periods used between interventions? Did the investigators analyze for carry-over or other period or sequencing effects with a cross-over design?
- **Did the study randomly assign patients to treatment groups using a method that was truly random (i.e., randomization)?**

- Was the study blinded to the extent possible (double-blinding preferred)? If blinded (single or double), was unblinding likely, e.g., side-effects, characteristic taste, smell or lab results for one treatment? Did the investigators determine the extent of unblinding and, if so, did it appear to have occurred?
- Were the drugs administered for a sufficient duration given the study's purpose/objective?
- If concurrent nonstudy medications were allowed, were any used that might affect either the study treatments or results? If so, did the study quantitate and compare their use to ensure it was similar between groups?
- Were adverse effects reported and statistically analyzed? Could they affect the clinical applicability of therapy?
- Was adherence to treatments and other study requirements measured (ideally using more than one method), with the findings reported? If yes, could adherence differences have led to any differential results seen? If no, how could this affect the study findings?

Outcome measures (refer to Chapter 7)

- Were the primary and secondary outcome measures clearly defined and appropriate for the study objectives?
- Were any needed investigator and/or patient training provided (e.g., for use of devices, diaries) and standardized methods used (e.g., clearly defined protocols, procedures, definitions) to ensure that outcome measures were appropriately carried out and interpreted, especially across multicenter study sites?
- Was the timing of outcome measurements appropriate and of adequate frequency and duration?
- Were the different patient groups handled similarly except for the treatments studied?

Statistical methods (refer to Chapter 8)

- Were appropriate statistical tests used for all outcome measures, taking into account the level/scale of the data, the number of comparisons, and whether the data represent paired or unpaired samples?
- Did any correlation (r) values reported represent strong or clinically important associations?

Results (refer to Chapter 9)

- Were any significant differences apparent among groups at baseline that could influence study results?

- Were any important baseline comparisons of factors that might affect the study findings overlooked?
- Were there factors (e.g., study setting, diet, other confounding variables) besides the treatments used that could have affected the results observed? If so, were they accounted or controlled for in the study?
- Was the number of patients accounted for at each step of the study and was it clear how many patients were in each analysis?
- Could the reasons for drop-out affect the clinical usefulness of therapy? Did the data-handling method used significantly affect the interpretation of study findings?
- Was power appropriate for all primary and secondary outcome analyses, considering drop-outs and possible use of the exclusion of subjects data-handling method? Was power sufficient for any subgroup analyses performed?
- If power was insufficient or not reported for an outcome measure of interest, did type II error appear likely for nonstatistically significant findings?
- If the data appeared skewed, was the median included as a measure of central tendency?
- Were the measure(s) of variability used appropriate (e.g., use of SD and not SEM, use of interquartile range for skewed data) and sufficient (e.g., confidence intervals included)?
- Were findings statistically significant and, if so, were they large enough to be clinically significant (important)?
- Was a number needed to treat (NNT) provided, where appropriate, to help determine the clinical applicability of the findings?

Discussion/conclusions (refer to Chapter 9)

- Were the results (positive and negative) interpreted appropriately by the authors?
- Did the authors adequately explain key study limitations and any discrepancies from other similar studies?
- Were authors' conclusions consistent with the results and study limitations and extrapolated appropriately?

Overall summary/assessment

- What were the *important* weaknesses of the study and, given these, what key points/findings should be taken away from the study?
- Could any study limitations or design weaknesses reduce internal validity, thereby affecting its external validity (generalizability)?

- In light of the study and its findings, what should be the role of the study drug(s) in current therapy or the clinical practice implications of the therapy studied?
- Is any further research needed and, if so, what is needed and how should this research be conducted?

These questions were adapted from the paper by Blommel and Abate (2007).

11
Equivalence and noninferiority studies

Learning objectives

Upon completion of this chapter, you should be able to:

- define an equivalence study and a noninferiority study
- describe the main differences between equivalence and noninferiority studies and the usual "superiority" studies
- describe how to interpret the findings from equivalence and noninferiority studies
- describe important potential problems with noninferiority studies.

Overview

The previous chapters focused on the clinical experimental trials conducted most frequently, namely, "superiority" studies. These studies are generally designed to determine if a new treatment is better than the control, which can be a placebo or active therapy. As discussed in Chapter 9, the findings from these studies are either that the treatment is significantly different than the control (i.e., reject the null hypothesis of no difference; thus, one therapy is superior to the other), or the study treatment is not different than the control (i.e., accept the null hypothesis of no difference in therapies). However, if the null hypothesis is accepted, it does *not* mean that the treatments are definitely the same – one can only conclude that no significant difference was found (refer to Chapter 9 for more details). In addition to the usual clinical "superiority" trials, however, there are two other types of controlled experimental studies that clinicians should be aware of. These include *equivalence studies* and *noninferiority studies*. In particular, noninferiority studies are being reported with increasing frequency in the literature.

With both of these study designs, an active control is used (*not* a placebo). These studies are designed to determine if a new therapy is equivalent to (no better or no worse than) an active control therapy (equivalence study), or at least as efficacious as (no worse than) an active control (noninferiority study). For more detailed discussion of the concepts reviewed in this chapter, the reader is referred to reviews by Piaggio et al. (2006), Leon (2011), and Lesaffre (2008).

When reviewing the appropriateness of equivalence and noninferiority studies, similar questions should be asked as outlined in Chapter 10 for "superiority" studies. Regardless of the type of experimental study, the journal the study was published in should be of sound quality and the study should be appropriately designed and performed, with the data properly analyzed to ensure that the stated objective was adequately addressed. However, there are some differences from "superiority" studies, particularly with the noninferiority design, that also need to be considered when evaluating their quality. This chapter provides a brief review of equivalence and noninferiority studies.

Equivalence study

In an *equivalence study*, the investigators are not concerned with determining if a new treatment is better than another therapy. Rather, they simply want to show that the treatments are not significantly different from each other (i.e., they are equivalent). The easiest way to think of an equivalence study is to think of a study that is trying to show that the bioavailability of a new generic drug is similar (equivalent) to the bioavailability of the control (brand-name) drug. The generic drug's manufacturers are usually not out to prove that the new generic drug has better bioavailability than the brand-name product; they would be happy with showing that it has equivalent bioavailability. Thus, their objective might be to demonstrate that the generic drug has similar bioavailability to the brand-name product using an equivalence study design.

To show that the generic drug has equivalent bioavailability, the investigators will choose an "equivalence" margin or interval at the start of the study. For the bioequivalence example, this interval would be the range of bioavailability values considered equivalent to that of the brand-name product. If the bioavailability results for the generic drug fall within that range, the generic and brand drugs would be concluded to have equivalent bioavailability. Equivalence studies could also focus on outcomes other than bioequivalence, but they should not be used for clinical studies comparing therapy efficacy.

> *Equivalence study example*
>
> Let's say an equivalence study examines the bioavailability of new drug A compared to established drug B. The investigators identify at the start of their study an equivalence margin (expressed as a 95% confidence interval (CI)) of 80–125%. This means that drug A will be concluded to be equivalent to drug B if the bioavailability of drug A is anywhere from 80% to 125% that of drug B's bioavailability.

> Suppose the study finds that the bioavailability of new drug A = 92%, with a 95% CI = 88%–96%. Since this CI falls completely within the specified equivalence values of 80–125%, the drugs would be concluded to be bioequivalent. If the bioavailability of new drug A was outside that range (i.e., some or all of the reported CI values were either below 80% or above 125%), the conclusion would be that drug A is not bioequivalent to drug B.

Noninferiority study

Noninferiority studies are being reported increasingly in the literature. Although their use is not yet widespread, clinicians should be familiar with this type of study since the results are interpreted in a different manner from the usual "superiority" studies. A *noninferiority study* is used to determine if a treatment has *at least* the same efficacy as an established active control. This means that the study treatment could have better or the same efficacy as the active control, but not worse.

> *Why perform a noninferiority study?*
>
> A noninferiority study might be useful for:
>
> - determining if a new drug with fewer side-effects or other secondary benefits (e.g., raises "good" high-density lipoprotein cholesterol, costs less, easier administration, better patient adherence) is at least as efficacious as an established therapy. If the new drug is shown to be noninferior in efficacy to the established therapy, the new drug might be preferred due to the secondary benefits
> - establishing the efficacy of a new drug when there is an existing highly beneficial therapy and it is unethical or not possible to use a placebo control.

In a noninferiority study, the investigators would like to determine if a new therapy is at least as efficacious as, but not worse than, the active control. If the new (N) therapy efficacy is subtracted from the control (C) efficacy, N − C, or the reverse is done, C − N, one obtains the difference in efficacy between the two treatments. A key feature of a noninferiority study is that the investigators *must* specify in advance the "allowable" difference in efficacy for which N would still be considered at least as good as C.

Thus, a noninferiority trial will specify a certain efficacy difference (called the *noninferiority margin* or *threshold*) that the lower value of the treatment CI must not fall below (or the upper limit must not be above, depending on which value is being subtracted from the other) for N to be considered noninferior to C.

It is easiest to think of the noninferiority threshold as the largest *clinically acceptable* difference between groups that would still mean that the therapies are at least as good as each other. For example, a noninferiority study comparing the otitis media cure rates with new drug N to active control drug C might specify a noninferiority difference in cure rates of 8%. This would mean that the drug C cure rate minus the drug N cure rate should not exceed 8% in order for drug N to be considered noninferior to drug C. It should be noted here that, unlike the usual "superiority" studies, noninferiority studies often do not report P values (refer to Chapter 9 for a discussion of P values) but instead report a 95% CI for the efficacy rates. So, in order for drug N to be considered noninferior to drug C, a value of 8% (or greater) should not be in the CI range for the difference in efficacy between drugs N and C.

Worked example

A noninferiority study compared 12 weeks of therapy with new drug X to active control drug Z to treat rheumatoid arthritis. The study set a noninferiority margin of 5% for the difference in arthritis response rates between drug X and drug Z. Thus, drug X was considered to be noninferior to drug Z if the difference between their arthritis response rates (i.e., drug Z response rate minus drug X response rate) was not greater than 5%. After 12 weeks, the study found that the response rate with drug Z was 75.8% (95% CI = 72.6–79.0%) and the response rate for drug X was 76.0% (95% CI = 72.3–79.7%); difference between groups was −0.2% (95% CI = −4.3% to 4.7%).

Was noninferiority of drug X shown in this study?

Yes. The study found a between-treatment difference (drug Z efficacy minus drug X efficacy) in arthritis response rates of −0.2%. The reported CI for the treatment difference tells us, with 95% confidence, that the population difference in efficacy between drug Z and drug X will fall from 4.3% less, to up to 4.7% greater, with drug Z compared to drug X. Since the upper value in the CI, 4.7%, is less than the predefined noninferiority margin of 5% (and all the other CI values are likewise less than this cut-off), noninferiority was demonstrated.

Could the noninferiority limit ever be a negative number?

Yes. In this example, the investigators chose to subtract the new drug X efficacy from drug Z efficacy, i.e., drug Z − drug X. When subtracting the drugs in this order, when control drug Z has greater efficacy than drug X, the resulting efficacy difference would be positive. If the investigators had instead chosen to subtract drug Z efficacy from new drug X efficacy, i.e., drug X − drug Z, when specifying their noninferiority margin, greater efficacy of control drug Z would result in a negative inferiority margin. The greater the efficacy difference for drug Z compared to drug X, the larger the negative number will be.

So, when you read noninferiority studies, the specified inferiority margin might be positive or negative depending on which drug's efficacy value (e.g., the new study drug or the proven active control) is being subtracted from the other, and also on whether the outcome is a beneficial one (e.g., response rate) or an undesirable one (e.g., thrombosis development).

Null hypothesis and type I and II errors in noninferiority studies

Noninferiority studies can be tricky to interpret because the null hypothesis is different from that described for the usual "superiority" studies (see Chapter 6). A noninferiority study wishes to show that a new treatment is noninferior to the control therapy by defining an "allowable" clinically acceptable limit for any efficacy difference. In the worked example, this efficacy difference was set at 5%. If the efficacy difference between the drugs actually exceeded 5%, the conclusion would be that new drug X is inferior to active control Z. *The null hypothesis for a noninferiority study is that the new treatment is inferior to the active control.* The null hypothesis is accepted any time the defined noninferiority limit is equaled or exceeded; when this occurs, the new treatment is concluded to be inferior to the active control. When the difference in efficacy between drugs is less than the defined noninferiority limit, the null hypothesis is rejected and it is concluded that the new treatment is noninferior to the active control (i.e., the new treatment is at least as efficacious as the active control).

Interpreting P values in a noninferiority study

As discussed in Chapter 9, $P = 0.05$ is usually considered the cut-off for statistical significance for typical "superiority" studies. Noninferiority studies will often report a CI instead of a P value. However, when a noninferiority study reports a P value, $P = 0.025$ is generally used as the cut-off for statistical significance. When interpreting reported P values for efficacy in a noninferiority study, use the following as a guide unless otherwise specified by the study:

- If $P < 0.025$, reject the null hypothesis and conclude that the new treatment is noninferior (has at least similar efficacy) to the control.
- If $P \geq 0.025$, accept the null hypothesis and conclude that the new treatment is inferior to the active control.

A type I error is rejecting the null hypothesis when it is really true, and a type II error is when one fails to reject the null hypothesis when it is false and should be rejected (see Chapter 9). For a noninferiority study, a type I error is possible when $P < 0.025$ and a type II error is possible when $P \geq 0.025$. However, because the definition of a null hypothesis differs in a noninferiority study from that in the usual "superiority" study, a noninferiority study interprets type I and type II errors using its null hypothesis definition. Thus, in a noninferiority study, a type II error means that the study wrongly concluded that the new treatment was inferior to the control therapy, when it really was at least as efficacious as the control. A type I error means that the study wrongly concluded that the new treatment was noninferior to the control.

Key Points

The following points should be remembered about the null hypothesis and type I and type II errors in a noninferiority study:

- The null hypothesis is that the new treatment is inferior to the active control.
- *P* values are less commonly reported in noninferiority studies; a CI is usually reported.
- Type I error ("false positive") is *possible* when $P < 0.025$.
- A type I error occurs when the study incorrectly concludes noninferiority of the new treatment to the active control.
- Type II error ("false negative") is *possible* when $P \geq 0.025$.
- A type II error occurs when the study "misses" (i.e., incorrectly fails to conclude) a finding of noninferiority of a new treatment compared to the active control.
- Insufficient statistical power (usually from enrolling too few patients in the study) will favor a type I error (as opposed to a "superiority" study in which too low a power favors a type II error).

Not only do the definitions of type I and type II errors differ in a noninferiority study, but the occurrence of these errors also seems opposite to that in the "superiority" study. This means that, in a noninferiority study, a lack of power favors a type I error (wrongly concluding noninferiority), while lack of power in a "superiority" study favors a type II error (see Chapter 9).

Potential problems with noninferiority studies

Although there are some valid reasons for performing a noninferiority study, as previously indicated, there are also potential problems with this type of study design. Some of these problems include:

- *How large should the noninferiority margin be? How does one determine the size of the difference that would make a new drug "just a little inferior" but still clinically acceptable to a proven therapy?* These questions are difficult to answer. One does not want the new drug to be considered "noninferior" to proven therapy only because the noninferiority margin was set too large. Otherwise, the new drug may really be less clinically beneficial than the active control.

Thus, investigators will often set a relatively small noninferiority margin/threshold, as compared to the effect size used to determine sample

size in a "superiority" study. This results in the need for a larger study sample size for a noninferiority study, which might be difficult for the investigators to achieve.
- In a "superiority" study, if the study does not have sufficient power to show there is a real treatment effect, it wrongly concludes that there is no significant difference in efficacy between treatments. *With a noninferiority study, failure to detect a real treatment difference would likely result in an incorrect conclusion that the new drug is noninferior (i.e., has at least similar efficacy) to the control.* This could have detrimental effects in practice if the new, less efficacious drug is used instead of the active control.
- *What if the active control does not show the expected efficacy in a noninferiority study?* This might be due to the types of patients included. In this case, the study will likely find the new drug to be noninferior to the active control, but neither drug might actually have been efficacious.
- *A noninferiority study gives only the relative efficacy of a new drug compared to an active control.* One still does not know how the new drug would perform compared to a placebo control.
- *Pharmaceutical manufacturers might be tempted to conduct a noninferiority study to show that a "me too" drug, one with no real advantages in practice, has similar efficacy to an established therapy.*

A summary of key points and how to apply the information from this chapter to practice follow.

Key Points

Due to the differences and potential for confusion on the part of clinicians who read noninferiority versus "superiority" studies, the following should be clearly stated in a noninferiority study:

- the rationale for choosing that particular design as compared to a "superiority" study
- the noninferiority hypothesis
- the noninferiority margin/threshold along with how it should be interpreted
- how sample size was calculated; make sure the power was reported and was adequate
- why the active control used was selected for the study
- whether the active control showed the expected efficacy in the study
- how the results should be interpreted.

How to apply to practice

- Look carefully at the margin/threshold used in a noninferiority study and determine if it appears reasonable (i.e., not so large that a possibly inferior new drug might still be shown to be "noninferior").

- Consider whether the investigators really needed to use a noninferiority design – ask whether the new drug being studied has clinically important secondary advantages over the control drug.
- Be extra cautious about applying the results from a noninferiority study that appears to be comparing a "me too" drug to an established active control drug. Do there appear to be potential conflicts of interest for the study investigators that might have influenced their selection of this study design and the interpretation of its findings?

Self-assessment questions

Question 1
What is the importance of the noninferiority margin used in a noninferiority study? Briefly explain.

Question 2
What is the null hypothesis in a noninferiority study? Does it have the same meaning in a "superiority" study?

Question 3
What P value is usually used as the cut-off for statistical significance in a noninferiority study?

a. 0.01
b. 0.025
c. 0.05
d. 0.10

Question 4
Which of the following is more commonly reported in a noninferiority study for the efficacy results: P values or CIs?

Summary

This chapter provided definitions and a brief overview of two additional types of experimental trials that clinicians should be familiar with, equivalence and noninferiority studies. While the considerations for evaluating the quality of the more commonly conducted "superiority" studies would apply to equivalence and noninferiority studies as well, some important differences were described.

12

Practice guidelines, systematic reviews, and meta-analyses

Learning objectives

Upon completion of this chapter, you should be able to:
- identify the purposes of systematic reviews and meta-analyses
- describe a systematic review and how it is prepared
- describe a meta-analysis and how it is prepared
- explain the potential problems that can occur with a meta-analysis and how they can be avoided or minimized
- describe how clinical practice guidelines are best developed
- identify good sources for locating systematic reviews, meta-analyses, and practice guidelines.

Introduction

Practitioners should critically read and evaluate individual studies to identify whether, and how, the findings can be applied to practice. Determining how to apply the results to practice becomes more difficult, however, when a number of different studies address a similar topic but arrive at varying conclusions. Is there any type of publication that can help busy clinicians remain current with the literature and address clinically important questions? Yes, there are resources available that analyze and interpret the results from several studies (a body of literature) in a given subject area to provide guidance when making clinical decisions. These resources include clinical systematic reviews, a special type of systematic review called a meta-analysis, and practice guidelines. This chapter will briefly outline the features of these resources and how they can be located.

Systematic reviews

A *systematic review* is undertaken to answer or explore a specific research question by thoroughly examining the relevant literature in an area using a structured, rigorous process. This process involves:

- clearly identifying the review's objectives
- specifying inclusion/exclusion criteria for the studies to include

- thoroughly and comprehensively searching for studies that meet the eligibility criteria
- critically examining the strengths and limitations of the studies included and assessing the validity of their findings by using clear, explicit methods
- formulating conclusions based on the evidence presented.

Meta-analyses

A *meta-analysis* is a type of systematic review. In addition to using a structured, rigorous approach, it essentially takes a group of individual studies on a topic and statistically (quantitatively) combines the data from each study to calculate a pooled, overall outcome measure. The steps involved in performing a meta-analysis are analogous to those used for a clinical study, except instead of enrolling individual patients, a meta-analysis incorporates individual *studies*. The meta-analysis steps include: (1) identify a clinical problem; (2) develop eligibility criteria for the studies to include; (3) locate the studies; (4) extract the needed data from each study using independent (ideally blinded) observers; (5) establish a scale to rate the quality of acceptable studies; (6) use a standardized outcome measure to determine and chart (with 95% CI) the difference between the treatment and control groups for each study; and (7) statistically calculate the combined outcome measure for the meta-analysis. The combined outcome measure is usually reported as an effect size (defined here as the difference in the outcome between intervention and control groups divided by SD) for continuous-level data and as an OR or RR for categorical data.

A *cumulative meta-analysis* can be thought of as an ongoing meta-analysis. This means that each time another study is performed that is similar to the others included in a meta-analysis, the new study's results are statistically added. This generates a new, pooled outcome measure based on the use of a larger meta-analysis sample size. For clinical questions that different investigators are each actively researching, a cumulative meta-analysis might allow for statistically significant findings to be uncovered sooner than waiting for any one, sufficiently large study to be completed.

> *Why perform a meta-analysis?*
>
> Sometimes, a number of studies have already been performed to examine a specific clinical problem. If each of these studies was rather small, they might not have had the sample size (i.e., statistical power: refer to Chapter 9 for a discussion of power) needed to show a significant treatment effect. Or, their findings might disagree. When the meta-analysis combines the results from these studies to calculate a

combined, standardized outcome measure, the sample size is effectively increased (the patients from each individual study are now combined into a much larger overall sample in the meta-analysis). As a result, a meta-analysis can:

- increase statistical power compared to smaller, individual studies and, by doing this, might help to resolve disagreements among the individual studies
- potentially determine an answer to a clinical question much faster than it would take to complete an individual study large enough to have the needed power
- provide more precise (accurate) estimates of treatment effects
- answer questions not posed at the start of individual studies
- enable conclusions to be generalized to a more varied range of patients
- provide the most reliable treatment recommendation in the absence of a large, definitive clinical trial.

What types of topics do meta-analyses cover?

Meta-analyses can now be found on most clinical topics. A few examples include:

- comparative efficacy and safety of enoxaparin and heparin
- review of soy isoflavone supplements used for osteoporosis
- statin efficacy in men compared to women
- antidepressant efficacy in older patients.

Are there any potential problems with a meta-analysis? The answer is definitely yes. Because a meta-analysis is a type of "observational" study, it is subject to a number of possible biases. A summary of these problems and biases and ways to avoid or minimize them follows:

- Heterogeneity – the studies included might have important differences (e.g., patient population targeted, drug doses/dosage forms/administration, study design, outcome measures) that might make them inappropriate to combine in a meta-analysis.
  - Avoid by testing for heterogeneity. This can be done statistically, graphically, or by conducting a sensitivity analysis, in which the combined outcome measure is calculated with and without certain types of studies included (e.g., small versus large studies, those using lower versus higher doses). If there is no bias or heterogeneity, the pooled outcome measure should stay the same when certain studies are

excluded. A sensitivity analysis is an important method for identifying possible bias in a meta-analysis.
- Selection bias – bias in selecting only certain types of studies for inclusion.
  - Avoid by using predefined criteria for meta-analysis inclusion and blinded persons to code and extract data.
- Publication bias – can occur when meta-analysis only includes published studies: (1) published studies can be more likely to show significant results than non-published studies; and (2) small studies, or studies funded by pharmaceutical manufacturers, are more likely to be published if they show significant results.
  - Avoid by including unpublished studies to extent possible.
- English-language bias – can occur when a meta-analysis only includes English-language studies, since foreign-language studies published in English are more likely to show significant findings than those published in foreign-language journals.
  - Avoid by including foreign-language studies.
- Database bias – occurs when a meta-analysis relies only on databases, or when it uses only a very small number of databases, to locate studies for inclusion; databases are less likely to index journals from less-developed countries and those journals that are indexed are more likely to show significant findings.
  - Avoid by using a variety of databases and other resources to locate studies.
- Citation bias – occurs when a meta-analysis relies heavily on bibliographies of published studies for locating other studies to include, since these studies are more likely to show significant effects.
  - Avoid by using a variety of resources to locate studies.
- Multiple publication bias – occurs when results from the same study are published in more than one place and the meta-analysis includes duplicate results.
  - Avoid by excluding multiple publications.

Another technique for detecting biases (including publication bias, English-language bias, any biases related to small sample sizes) in a meta-analysis includes the *funnel plot*. A funnel plot graphs (plots) the odds ratio (standardized outcome measure) from each study identified for inclusion in the meta-analysis (x-axis) against its sample size (y-axis). The resulting graph should resemble an inverted symmetrical funnel if no sample size or publication-related bias is present. Why? As a study's sample size increases, its outcome measure would better reflect the actual (population) treatment effect. So, the outcome measures from the larger studies would be at the top and clustered around the calculated value for the overall, combined meta-analysis (mid-point of the graph), while the outcome measures from

Table 12.1 How to locate systematic reviews and meta-analyses

Name	Source
National Library of Medicine: Pubmed	http://www.ncbi.nlm.nih.gov/pubmed/ Filter: type of article: meta-analysis
The Cochrane Collaboration; The Cochrane Library (source of several databases, including the Cochrane Database of Systematic Reviews)	http://www.thecochranelibrary.com/view/0/index.html
Excerpta Medica: Embase	http://www.embase.com/ Indexes systematic reviews; subscription needed
National Health Service (NHS); National Institute for Health Research – Centre for Reviews and Dissemination	http://www.crd.york.ac.uk/crdweb/ Access to the Cochrane Library's Database of Abstracts of Reviews of Effects (DARE), NHS Economic Evaluation Database, and the Health Technology Assessment (HTA) database

the smaller studies (closer to the bottom of the graph) should be more widely scattered around the midpoint. Refer to Egger and Smith (1998) for an illustration of a funnel plot and to learn more information about several of the biases associated with meta-analyses. For an overview of meta-analysis principles and an example of a critique that applies to those principles, refer to Chinchilli (2007). See Table 12.1 for resources to use to locate systematic reviews and meta-analyses.

Practice guidelines

According to the Institute of Medicine (1990), clinical practice guidelines are "systematically developed statements to assist practitioner and patient decisions about appropriate healthcare for specific clinical circumstances." They provide recommendations for the diagnosis or treatment of specific medical conditions that can be broadly applied to the patient group they are targeting. Clinical practice guidelines are designed to:

- provide a concise summary of the expected outcomes for available therapies
- assist practitioners in making decisions to help their patients; they are *not* meant to be "blindly" followed without consideration of the circumstances affecting individual patients
- reduce unnecessary or inappropriate use of tests, procedures, or treatments
- reduce variability in standards of practice from region to region.

"Systematically developed" in the definition provided above is critical to the utility of practice guidelines; meaning:

- specified, rigorous methods should be used for their development
- methods used should be transparent – how they were developed and funded should be clearly stated
- potential conflicts of interest by developers should be minimized or avoided
- the composition of the development team should be multidisciplinary and include representatives of the population affected by the guideline
- systematic reviews should only be incorporated if they also meet rigorous standards for their preparation
- recommendations should be clearly and precisely articulated, and include their benefits versus risks, a summary of the quality and quantity of available evidence, how (if) opinions and experience were used in their development and any differences of opinion, and ratings of the level of confidence in the underlying evidence and strength of the recommendation
- they should undergo external review, dates should be provided for when each stage of guideline development occurred, and they should be regularly updated when needed.

(Institute of Medicine, accessed at: http://www.iom.edu/Reports/2011/Clinical-Practice-Guidelines-We-Can-Trust/Standards.aspx, October 2012).

In light of the rigorous manner in which clinical practice guidelines are developed, are they used frequently by physicians? The answer is, not to the extent that they could (or should) be. Refer to an editorial by Alpert (2010) for several postulated reasons for less than optimal guideline use. Since practice guidelines can be very helpful resources for practitioners, it is important to know how to locate them quickly and easily. Table 12.2 lists several sources for these guidelines.

Worked example

Example 12.1

You are looking for a systematic review examining the efficacy of antioxidant vitamins to prevent macular degeneration. You locate two reviews:

1. Review 1: authored by three ophthalmologists who described the use of Medline to locate relevant articles on this subject; they then described the strengths and weaknesses of each article and summarized the topic by providing their conclusions.
2. Review 2: authored by an ophthalmology clinic at a university; they stated their methods for conducting the searches for relevant articles in several databases, outlined the eligibility requirements for the studies included, prepared tables outlining the key features of each study included and rated the quality of each study using a validated rating scale, and rated the overall quality of the evidence upon which each of their conclusions was based.

Which best meets the definition of a systematic review?
Review 2. Structured and well-defined methods were used for each part of the process undertaken. Multiple databases were used to increase the likelihood of locating relevant studies and a validated scale was used for quality ratings.

Table 12.2 How to locate clinical practice guidelines

Name	Source
National Guideline Clearinghouse	http://www.ngc.gov/
National Library of Medicine: Pubmed	http://www.ncbi.nlm.nih.gov/pubmed/ Filter: type of article: practice guideline
American College of Physicians: American Society of Internal Medicine (ACP-ASIM)	http://www.acponline.org/clinical_information/guidelines/
National Comprehensive Cancer Network (NCCN) (limited to oncology)	http://www.nccn.org/index.asp
National Institutes of Health	http://www.nih.gov/ Check individual NIH institutes for relevant practice guidelines
US Department of Veterans Affairs	http://www.healthquality.va.gov/
eGuidelines (online edition of Guidelines)	http://www.eguidelines.co.uk/new_guidelines.php Available at no charge for registered doctors or members of a variety of professional organizations in the UK
National Health Service (NHS) Evidence; National Institute for Health and Clinical Excellence (NICE)	https://www.evidence.nhs.uk/

Worked example

Example 12.2

You locate a meta-analysis that examines whether garlic can lower serum cholesterol. Specific eligibility criteria were specified for the studies to include in the meta-analysis and two blinded individuals reviewed the studies for data extraction. Six studies were used that each measured the effects of garlic on low- and high-density lipoprotein cholesterol concentrations. These studies were located by conducting a Medline search for relevant literature published in English. A sensitivity analysis showed that the studies included appeared to be homogeneous.

Any biases apparent in this meta-analysis?
Yes. English-language bias (no attempt was made to locate foreign-language studies), publication bias (only published studies were located with a Medline search), and database bias (only one database was used) are possible. It is not known if the other biases were potential concerns based upon the information provided.

Were any steps used to minimize potential problems?
Yes. The meta-analysis minimized possible selection bias (by specifying criteria for study inclusion and blinding the data extractors) and examined the studies included for heterogeneity or other possible biases (use of sensitivity analysis).

A summary of key points and how to apply the information from this chapter to practice follow.

Key Points

- Systematic reviews, meta-analyses, and practice guidelines can help keep clinicians up to date and provide useful guidance for making clinical decisions.
- These resources should state and use clear, specific, and rigorous methods for their development.
- There are a number of online resources available for easily locating all three types of publications on topics or clinical questions of interest.
- While providing very useful information, meta-analyses are also subject to several potential problems and biases that can affect their quality, including: (1) heterogeneity among the studies included, and (2) selection, publication, English-language, database, citation, and multiple publication biases.

How to apply to practice

- Look for a systematic review, meta-analysis, or clinical practice guideline when you want to remain updated on a certain topic or when you need to locate information about a specific clinical or research question.
- Critically read and evaluate meta-analyses to determine if they were conducted in a well-defined, rigorous manner. Well-conducted meta-analyses should generate funnel plots and perform sensitivity analyses and tests for heterogeneity to check for "outlying" studies and potential biases in the studies included.
- Use well-developed clinical practice guidelines to help inform and guide patient care decisions.

Self-assessment questions

Question 1
Which of the following are correct regarding the definition of a meta-analysis or its purposes?
a. A meta-analysis uses formal statistical techniques as a key part of its analysis.
b. A meta-analysis can help generalize conclusions from individual studies to a more varied range of patients.
c. A meta-analysis combines the results of several independent clinical studies.
d. A meta-analysis is done to provide a detailed analysis of a single clinical trial.
e. A meta-analysis can provide answers to questions that are not posed at the start of individual clinical trials.

Question 2
Briefly define a "cumulative" meta-analysis.

Question 3
Explain whether the following statement is accurate: "A meta-analysis has an advantage over other types of studies in that it is relatively free of potential biases."

Question 4
The investigators who performed a meta-analysis state: "We used Medline to locate the studies that were included in our analysis. Several different search terms were used when searching Medline to thoroughly locate all relevant articles, including foreign-language studies." Based on this statement, is it likely that publication bias was present in this meta-analysis?

Question 5
A meta-analysis of the use of antibiotics to treat acute otitis media included 18 studies and reported that "sensitivity analyses were performed to assess the influence of trial design on the pooled OR reported." What do sensitivity analyses refer to here?

Question 6
Is there a database that can be used to locate clinical practice guidelines for a wide variety of subjects?

Summary

There are a number of available resources that can assist busy clinicians in finding the answers to clinical questions in an efficient and effective manner. These resources include the use of systematic reviews, of which meta-analyses are a particular type of systematic review, and clinical practice guidelines. This chapter reviewed the definition and features of systematic reviews, meta-analyses, and practice guidelines and provided several sources to use for locating this type of information.

Answers to self-assessment questions

Chapter 1

Question 1

An important concern with using a systematic approach is ensuring that the process itself does not become fragmented. If each step is approached as an independent task or lacks a logical and concise flow, the process will be choppy and inefficient. While each step in a systematic approach is important, no step should stand alone without consideration of the other involved steps. The steps of any approach should be used as a guide to ensure the response is accurate and thorough, but each step should build on previous ones.

Question 2

The first question asked should determine if a specific patient is involved. If so, additional questions should be asked such as: Is the patient currently pregnant or planning on becoming pregnant? What is the lisinopril being used to treat (e.g., hypertension, proteinuria, heart failure)? What trimester is the patient in and when during the pregnancy did she receive lisinopril? How long did the patient receive lisinopril and at what dose?

All of these questions help to clarify the clinical scenario in the question posed and allow for the provision of specific information to meet the needs of the requester.

Question 3

Did this patient experience a hypersensitivity reaction while using insulin? What other medications was the patient taking when the hypersensitivity reaction occurred? What type of insulin is the patient taking? How long has the patient been taking the insulin and any other medications? What dose of insulin and any other medications is the patient taking? What was the temporal relationship with insulin and any other medications to the hypersensitivity reaction? What were the characteristics of the hypersensitivity reaction? Has the patient ever experienced a similar type of reaction previously?

All of these questions can help to refine the clinical scenario in the question posed and might allow the requester to focus the direction of the response better.

Question 4

Limitations to consider when assessing tertiary resources include the currency of the information, the author's expertise in the subject area, the use and citation of references, and the presence of bias or other errors. Remember that the information found in tertiary resources can be outdated, due to a possibly long lag time when publishing print resources or due to inconsistent updating of online resources. The authors should have sufficient experience in the topic presented and the citation of references lends credibility to the information found in tertiary resources. At least two credible tertiary resources should be used to verify information found.

Question 5

It is an ethical responsibility of the responder to provide both the advantages and disadvantages of recommendations made as well as the limitations of the information found when providing responses to questions. The information retrieved should be analyzed and critiqued through the use of proper evaluation skills. The answer should include the strengths and limitations of the evidence used to prepare recommendations, which will allow the requester to understand fully the rationale behind them.

Question 6

Information requests and needs should be handled using an evidence-based approach. The original question should be restated if needed to include pertinent background information. Appropriate resources should be used to gather relevant information, and the response should detail the information found in a logical and concise manner. The response should address all important aspects of the actual question. Ideas presented should flow in a rational, logical manner that is easily followed and understood. The response should end with a brief conclusion and summary of important points.

Question 7

Follow-up allows the responder to assess if the information provided truly met a requester's needs. It also provides an opportunity to assess if there are additional questions or if any further information is needed. Follow-up also involves the provision of new information that might become available after the original response was delivered.

Chapter 2

Question 1

Since paroxetine does not have FDA-approved labeling for treating hot flashes, tertiary references that provide unlabeled or off-label uses of drugs should be consulted. Two of the general information sources that would be particularly suited to find this information are *Micromedex's Drugdex* and *AHFS Drug Information*. *Drugdex* would also provide a fully referenced summary of any data from the primary medical literature regarding this off-label use of paroxetine, while *AHFS Drug Information* would include references only in the electronic version.

Question 2

Many tertiary resources will provide the pregnancy risk categories of A, B, C, D, or X. However, these ratings do not provide specific details regarding the actual consequences of taking the drug during pregnancy. *Drugs in Pregnancy and Lactation* and the *Reprorisk* system in *Micromedex* provide information beyond a simple pregnancy risk rating. Both of these resources include summaries of available literature that describe the specific consequences of taking a specific drug during pregnancy. Both resources are fully referenced, which allows the clinician to evaluate the data sources and consult them for further discussion if needed.

Question 3

Multiple resources may provide a risk assessment or risk rating when taking topiramate during lactation. However, similar to the pregnancy risk ratings, lactation risk ratings may not provide sufficient detail about the consequences of taking the drug while breastfeeding. *Drugs in Pregnancy and Lactation*, *Medications and Mother's Milk*, and *Lact-Med* should be consulted to review the available literature related to topiramate use while breastfeeding and provide the patient with information that she can share with her doctor when discussing the situation.

Question 4

The *Natural Medicines Comprehensive Database* and *Natural Standard* should both provide information regarding the uses and relative efficacy of cinnamon, based on currently available information, when used to control blood glucose levels in diabetes. The *Natural Standard* presents literature supporting the relative efficacy in a tabular format that allows for quick analysis of the strengths and limitations of the primary sources used to determine efficacy.

Question 5

Finding recipes for compounded products can be challenging. Some resources to consult first when trying to locate a compounding recipe include: *Extemporaneous Formulations for Pediatric, Geriatric, and Special Needs Patients*; *Pediatric Drug Formulations*; and *Trissel's Stability of Compounded Formulations*. Each of these references contains formulas for compounding many different products. These references will also provide stability and beyond-use dating information for the compounding formulations included. It should be noted that, although the title of *Pediatric Drug Formulations* is focused on pediatric patients, the formulations could be used for patients of any age.

Question 6

Resources are needed that provide international sources for medications. *Index Nominum* and *Martindale* would be useful resources to locate this type of information. Both would allow a clinician to find basic information about international drug products and cross-reference that product to a domestic equivalent if available. If more detailed or comprehensive information about the medication is needed, *Martindale* would be the best reference to use since *Index Nominum* primarily serves as an index of products and does not provide therapeutic information (other than a listing of the therapeutic category) for these products.

Chapter 3

Question 1

Key terms should include any topics necessary to produce relevant results from a secondary resource. The key terms that should be included in this question are ketorolac, meperidine, diclofenac, pain, and kidney stones.

Question 2

The first step would be to locate the appropriate index terminology for the specific secondary resource being used. Each resource may have different terminology for the key terms identified. If using PubMed, the MeSH database should be used to identify the following MeSH terms that most appropriately match the key terms identified:

- the drug terms would use the generic drug names
- the kidney stone term as used in this question would match best with the MeSH term *urolithiasis*
- the pain term as used in this question would match best with the MeSH term *renal colic*

These MeSH terms were chosen based on the definitions provided in the MeSH database. It is important to check the specific definitions provided for each MeSH term to ensure the term matches exactly with the question topic. An additional note for drug terms is that it can be useful to search PubMed by simply using the generic drug name. Finally, once the appropriate MeSH terms have been found, the proper way to combine these terms would be "ketorolac AND (meperidine OR diclofenac) AND (urolithiasis [MeSH] OR renal colic [MeSH])". It is important to use the parentheses around the terms combined with the OR Boolean operator to ensure proper grouping of the subjects in the search string.

Question 3

Referring to the original question, the physician was interested in clinical trials evaluating the use of the drugs. Thus, in the above search any results that were not clinical trials (e.g., review article, case studies) would be irrelevant. By placing a "clinical trials" limit on the search, the results would eliminate the irrelevant publication types such as review articles. Use such limits cautiously, however, because what may be deemed irrelevant at first glance may actually provide useful information.

Question 4

This question focuses on a pharmacy-related issue, drug stability. IPA would be a good resource to use for this question since it indexes many pharmacy-specific journals and its scope of information includes drug stability. IDIS can be another appropriate choice since its scope of information also focuses on pharmacy and it has a specific descriptor used to index articles that provide stability information. It would be important also to use other secondary resources such as PubMed or Embase if information could not be located using the pharmacy-focused resources.

Question 5

Insulin and allergy are the key terms in this question.

Question 6

The search terms could be grouped as follows: "insulin AND (allergy OR allergic OR hypersensitivity OR allergies)." Even though the question only has two main key terms, the term "allergy" has synonyms that could also be searched. To ensure a complete search, multiple synonyms for allergy should be used in the search. The Boolean operator OR allows for the inclusion of the synonyms easily in the search. Again, parentheses should be used to group the terms being combined with the OR operator. It should be noted that it would be appropriate to use the controlled vocabulary term, if available, for "insulin" and "allergy" to retrieve the most specific results.

Question 7

Yes. Truncation would allow any number of word endings to be searched at one time. In this example, assuming an asterisk is the truncation symbol used by the secondary resource being accessed, the appropriate truncated word for allergy would be "allerg*." This would allow allergy, allergic, and allergies to be searched at one time.

Chapter 4

Question 1

The patient should be counseled that the HONcode symbol is used to denote websites that comply with a set of standards regarding how the information is presented. It helps ensure that the information is transparent, meaning that it is clear where the information came from and how it is being used on the website. However, the HONcode does not mean that the information itself was reviewed for accuracy and completeness. You should offer to review the information with the patient and assess the quality, accuracy, and completeness of the information she found. If the information is deemed to be of high quality, you should also assist the patient in applying the information found to her diabetes and its management.

Question 2

Remember that a date when last updated does not mean that *all* of the information was reviewed and updated at that time. Other items that should be reviewed to determine the currency of information are the references used and comparing the information to recent work and knowledge in the field.

Question 3

An internet search can be useful for looking up topics that are too recent to be included in other resources or published medical literature, the ingredients in marketed over-the-counter or multi-ingredient herbal and supplement products, and questions from patients who have heard or read their information in news reports.

Question 4

Healthfinder and MedlinePlus are excellent internet sources that will provide patients with information about a variety of topics. Each of these sites serves as an information gateway by providing links to selected high-quality health-related websites.

Question 5

The Centers for Disease Control and Prevention (CDC) website (www.cdc.gov) has excellent and comprehensive information for travelers to other countries, including Brazil. This website details the exact immunizations or prophylactic medications a traveler needs prior to departing for a certain country or region, as well as medications needed during or after the trip and any other precautions that should be taken (e.g., water sterilization, use of mosquito repellent).

Question 6

The FDA website (www.fda.gov) provides access to the electronic "Orange Book." This resource would provide therapeutic equivalence ratings for the Unithroid brand and the specific generic levothyroxine that the pharmacy stocks and would inform you if the generic could be legally substituted for the Unithroid.

Question 7

There are two websites that might be useful for finding information about this clinical trial. The National Cancer Institute's website (www.cancer.gov) has a link to clinical trial information through the PDQ cancer clinical trials registry. This is a fairly comprehensive registry specifically for cancer-related clinical trials. Another source that may provide information about this topic is the ClinicalTrials.gov website. The clinical trials in this database can be searched by topic, so prostate cancer can be entered as the search term.

Question 8

A review of the USP Verified website confirms that the one product does indeed carry the USP Verified seal. You explain to the patient that this seal ensures that the contents of the capsules meet the specifications on the label and have been shown to be pure and appropriately manufactured. Thus, the USP Verified product appears to be the better choice. Without this seal, there is no guarantee that the supplement contains the correct amount of ingredients as listed on the product labeling and is pure and free of contaminants.

Chapter 5

Question 1

Cohort (prospective, concurrent) design. Subjects were initially enrolled based upon whether or not they had the exposure (daily NSAID use). Medical records were used to identify these individuals at the start, but the men in both groups were actively followed prospectively for the next 5 years to compare the extent to which they developed the outcome (BPH). Since this is an observational study design, subject to more potential weaknesses and biases than an experimental study, the results from this study cannot definitely prove that regular NSAID use would delay BPH development.

Question 2

Prospective cohort design: a group of patients is identified with the drug use/exposure of interest (taking statin drugs for elevated cholesterol concentrations), along with another group of patients with similar characteristics (e.g., comparable ages, gender proportions, cholesterol concentrations) but without the drug use/exposure of interest (not taking statin drugs). Both groups of patients would then be followed into the future for a specified amount of time with their blood glucose concentrations periodically monitored. At the end of the follow-up period, the number of patients who developed the outcome (type 2 diabetes) during the study would be compared in both groups to determine if there is a significant difference.

Case-control design: a group of patients with the outcome (cases: type 2 diabetes) is identified along with another group of similar patients (e.g., comparable ages, gender proportions, cholesterol concentrations) but without the outcome (controls: without type 2 diabetes). Patients in both groups would be interviewed and/or medical records examined to determine their history of statin use. The number of cases and controls who used statins would be compared to determine if there is a significant difference.

Question 3

This is an experimental design because actual intervention was used by the investigators, i.e., the administration to patients of garlic tablets or placebos.

Question 4

Cross-sectional design. All study subjects were given a survey to determine their knowledge of both acetaminophen and its toxicity (the "exposure") and their ability to identify acetaminophen-containing products (the outcome) at one point in time. The survey results were analyzed to identify individuals with greater knowledge and those with less knowledge about acetaminophen, and to determine if there was an association between this knowledge and the ability to identify acetaminophen-containing products.

The cross-sectional design is not strong; it is one of the weakest designs. With this design, it is difficult to determine what came first, the ability to identify the products or the knowledge about acetaminophen. It also cannot be determined if the subjects recruited from the Boston clinic are truly representative of the rest of the population.

Question 5

A controlled experimental study is less subject to selection bias than observational case-control and cohort studies. With a controlled experimental design, patients are enrolled based upon specified criteria and the patients are then assigned to receive either the control or treatment groups. Thus, the patients in these groups would likely be similar to each other. In contrast, the patients in a case-control study (cases versus controls) or cohort study (study group versus comparison group) differ from each other from the start in that they either have or do not have the outcome (case-control design) or have or do not have the exposure (cohort design). The investigators have to select (enroll) patients in the study groups who are similar with regard to other important characteristics that might affect the findings; this is difficult to accomplish and could lead to selection bias (differences in the patients resulting from how they were selected for study inclusion).

Chapter 6

Question 1

Yes, there are potential conflicts of interest for this study's investigators. They include owning stock in the company (Ortho-NcNeil Neurologics) that manufactures galantamine, one of the investigators works for the manufacturer, and the funding to conduct the study came from the manufacturer. Supplying the study drug and placebo is not considered to be a conflict of interest. Keep in mind that simply because potential competing interests exist does not mean that the study is of poor quality. It does mean that extra care should be taken when reading the study to identify any flaws or biases that might unnecessarily favor the galantamine.

The null hypothesis for this study is: "there is no difference between galantamine and placebo on quality of life in Alzheimer's disease patients."

Question 2

The null hypothesis would be: "there is no difference between amoxicillin therapy given for 3 days and amoxicillin given for 8 days for hospital-acquired pneumonia in adults."

Question 3

"Peer-reviewed" here is referring to the journal's use of the peer review process for the articles it publishes. This means that, following review by editors (listed on the journal's website), a manuscript that meets the journal's goals would be sent to a small number of individuals with expertise in the subject area to review and critique the paper, make suggestions for change, and offer a recommendation about whether it should be published. The peer reviewers' comments are then reviewed by the editor(s), who make the final decision about publication.

Question 4

This hypothesis would be one-tailed since it is focused on studying change in only one direction, improving sleep quality. A two-tailed hypothesis would not give a direction of change, but would rather state that the researchers feel that melatonin will have an effect on sleep quality. If stated as a study objective (not a hypothesis), the study would indicate that they are studying the effect of melatonin on sleep quality or if melatonin will affect sleep quality.

Question 5

No, potential conflicts of interest are not apparent based on the information provided. The investigators are from an academic institution and the grant was funded by a federal government agency that should be unbiased with regard to whether or not the involved drugs will be efficacious. No other competing interests are stated.

It appears that peer review was likely from looking at the dates the manuscript was received and accepted. A 4-month time span was involved, sufficient time for peer review to have been performed. This can always be verified by looking up information about the journal online or by checking print issues for peer review information.

Question 6

If you notice something unusual about a study, or a design feature that makes you question why the study was performed that way, always consider whether possible conflicts of interest exist that might have influenced the investigators' objectivity. In this example, note that one of the authors works for Abbott Labs (manufacturer of drug A). Thus, the manufacturer has a vested interest in this study and its findings and this could be the reason for the selection of comparison drug B. From the manufacturer's perspective, even if drug A is found to be equally efficacious as drug B, the problems with drug B (cost, adverse effects) would likely result in clinicians favoring the use of drug A.

Chapter 7

Question 1

External validity.

Question 2

"Single-blind" means that the patients (most likely) are not aware of whether they are taking T or S but the investigators know the drug each patient was assigned. "Double-blind" is preferred to reduce the risk of bias. "Randomized" means that each patient had an equal chance of being assigned to receive either T or S therapy. Randomization is very important to minimize the likelihood that the patients in the study groups are different with regard to factors (both known and unknown) that might influence the results. Studies should state the method used for randomization so the reader can determine if it was appropriately performed. Since randomization does not guarantee that the patients in the study groups will be comparable to each other, the investigators should still compare postrandomization baseline characteristics between groups to see if any significant differences are present due to chance.

The use of concurrent NSAIDS could be a problem since these drugs can help relieve headache pain. If several doses are taken, the NSAIDS could reduce migraine pain severity and interfere with the patient's assessment of pain relief from T or S therapy. It is important in this study that the investigators quantitate and compare the use of NSAIDS in both treatment groups. If one group takes many more NSAIDS than the other and reports greater pain relief, it will be difficult to separate the NSAID effect from the drug treatment effect.

Question 3

"Open-label" means that the study was nonblinded – both investigators and patients were aware of the treatment the patients were receiving. An open-label study is subject to a significant amount of bias that could influence the study's findings. Since tinnitus is a subjective symptom experienced by patients, the patients' ratings of tinnitus might be influenced if they believed their assigned treatment was going to be efficacious. The investigators might also skew the findings by how and what they discuss with patients during study visits if they feel that one drug is going to more efficacious than another. Although an open-label study can provide interesting preliminary findings, the results should be confirmed in subsequent double-blind studies.

Question 4

No. The patients appeared to be enrolled into the study using a nonrandom convenience method – attending a pain clinic associated with a particular university. There is no indication that the 150 patients were randomly sampled from among *all* the patients who were seen at the three pain clinics. It is also not clear that *every* patient who met the eligibility requirements and who showed up at clinics between the dates listed was enrolled into the study. Since random assignment to treatment groups was used, a nonrandom sampling method is acceptable for clinical trials such as this one.

Question 5

Internal validity. Internal validity needs to be present before a study can have external validity (the ability to generalize its findings outside of the study).

Question 6

Sensitivity. A test that is sensitive means that it is able to detect the presence of a condition or characteristic when present. In this example, if joint erosion or joint space narrowing occurs, even to a small degree, the X-ray is able to detect it.

Question 7

Cross-over. With this design, patients are initially assigned to receive one of the study treatments for a certain length of time, then they are switched over to receive the alternative therapy for the next treatment period. With this design, it is important to use a wash-out period (time between treatments when no drug is administered) to allow the effects from the first treatment to disappear before the next treatment is given. This study did not use a wash-out period.

The parallel design is preferred over a cross-over study. Even when a wash-out period is used in a cross-over study, there might be some carry-over effects from the first treatment. Also, the longer study duration needed for a cross-over design compared to a parallel design might influence the medical condition being evaluated or otherwise affect the study results.

Question 8

Yes. The rating scale appears to lack validity and reliability. A number of psoriasis rating scales have been used and validated in practice, including the standardized Psoriasis Area and Severity Index (PASI). Refer to study by Robinson et al. (2012) for more information about psoriasis assessment. For any rating scales used in a study, clear definitions for each rating level should be provided to all study investigators across the study sites to help ensure consistency of the results. For example, the extent to which redness, thickness, and scaling were present as well as the percentage of skin area involved are important considerations for evaluating psoriasis and should have been specified for each rating scale level.

Chapter 8

Question 1

Chi-square test (assuming no cell has an expected frequency <5); otherwise Fisher's exact test – proportion of patients experiencing emesis represents nominal-level data; two independent (unpaired) groups involved.

Question 2

Repeated-measures ANOVA – repeated weekly measures of hemoglobin concentrations (continuous data; meets parametric test assumptions) in unpaired groups.

Question 3

Spearman rank-order r – measures linear association (correlation) between two variables; one of the variables is ordinal level (ranked overall symptom scores).

Question 4

One-way ANOVA – three independent (unpaired) groups, body weights (continuous data; meets parametric test assumptions).

Question 5

Paired t-test – two comparisons (paired – before and after in same patients) of body weight (continuous data; meets parametric test assumptions).

Question 6

Wilcoxon signed-rank test – two groups (paired, cross-over design with same patients in both group comparisons), serum glucose concentrations (continuous data not normally distributed so cannot use the parametric paired t-test).

Question 7

McNemar test – dizziness development (nominal-level data) in paired groups (cross-over design with same patients in both group comparisons).

Question 8

Mann–Whitney U test – two independent (unpaired) groups, VAS scores (VAS scales are measures in which patients indicate on a line their extent of pain, and the distance from the end of the line to the mark is measured using a ruler, representing continuous-level data), not normally distributed so cannot use the parametric unpaired t-test.

Question 9

Two-way ANOVA – three independent (unpaired) groups; BMD scores (continuous data; meets parametric test assumptions); the effects of two independent variables (type of treatment and baseline BMD) or one dependent variable (BMD sure after treatment) are being analyzed.

Question 10

There is an inverse linear association between quality of life ratings and extent of joint erosion, or, as the extent of joint erosion increases, the quality of life ratings decrease.

Question 11

Yes. The Spearman rank r is used to determine the linear correlation between two variables when one or both are either ordinal-level or continuous-level but not normally distributed. The quality of life rating is ordinal-level (ranked data) and joint erosion measured in millimeters is continuous-level, which might or might not be normally distributed.

Question 12

It is considered to be moderate to strong since it is in the range of 0.5–1, but close to the lower end of 0.5.

Question 13

The value of r^2, the coefficient of determination, represents the amount/proportion of variation in one variable that can be explained by changes in the other variable. In this example, $r^2 = 0.325$, which means that 32.5% of the variability in the arthritis quality of life ratings can be explained by the presence of the joint erosion. This indicates that other unexplained factors are responsible for most (67.5%) of the variability in quality of life ratings.

Question 14

Yes. ANOVA is used to compare continuous-level data among three or more groups for which parametric test assumptions apply. If this study finds a statistically significant ANOVA, it would mean that at least one of the group comparisons, placebo versus phenytoin 0.5%, placebo versus phenytoin 1%, or phenytoin 0.5% versus phenytoin 1%, is significantly different. However, it would not indicate which. To identify the between-group differences that are significantly different, one of a number of "post hoc" tests (which means looking at the data after the study is completed and, in this situation, after an ANOVA is performed) is used. Several post hoc tests were identified earlier in this chapter; the Fisher least significant difference (LSD) test is one of those used following a statistically significant ANOVA. If ANOVA was not found to be statistically significant, the post hoc test would not be needed.

Chapter 9

Question 1

b. Median

Question 2

The substantial difference between the median and mean concentrations indicates that the data are skewed (non-normally distributed). Most likely there are a small number of very large concentrations (outliers) that affected the mean value. The median would better show the central tendency (i.e., "typical" value) of the data set.

The IQR should be used with the median to show the variability or spread of the individual concentrations around the median value. The IQR provides the range of values between the 25th and 75th percentiles, which shows the spread of the middle 50% of the individual patients' values. The SD is best *not* used with skewed data; one can only conclude that ~68% of the individual study values will be within ±1 SD when the data are normally or near normally distributed.

Question 3

a is correct. The data appear skewed since the mean and median concentrations are different, and only the median would be unaffected by a few outliers. Medians can be reported for continuous-level data so answer b is incorrect. No measure of variability (e.g., SD) was reported with the mean value in either study, so there is no way of knowing the variability of individual patient concentrations. Thus, answers c and d are also incorrect.

Question 4

The bold statement is a 95% CI. It indicates that one is 95% confident that, in the population, the reduction in proteinuria in diabetic patients with the use of an ACEI and a low-salt diet would range from 32% to 66%. The population would be those individuals outside the study who meet the study's inclusion/exclusion criteria.

Question 5

a. The statistically significant P values would be those values that are less than 0.05. In this example they would be: 0.042, 0.04, 0.036, 0.021.
b. Power = $1 - \beta = 1 - 0.19 = 0.81 = 81\%$. The power should be 80% or greater so this power is acceptable.
c. Yes. Type II error, which is concluding that there is no treatment effect present and any difference is a chance finding (i.e., false negative), is *possible* any time $P \geq 0.05$ (the null hypothesis is accepted). For the L versus E comparison, $P = 0.06$ so it was not statistically significant and the null hypothesis of no treatment difference is accepted. The 8% difference seen between L and E is concluded to have resulted from chance. So, a type II error is possible.

The likelihood of a type II error is given by beta, which in this example was 0.19 (or 19%) for the comparison in SCr concentrations. An acceptable beta is 0.2 (20%) or less, meaning that when we conclude there is no statistically significant treatment difference, we are willing to accept that this conclusion might be wrong (a type II error) 20% of the time (or less). So, even though there is a ~19% risk of type II error for the L versus E comparison, it is considered acceptably low.

Power provides the likelihood of *not* making a type II error, which in this example would be $1 - \beta$, or $1 - 0.19 = 0.81$ or 81%. A power of at least 80% is acceptable.

d. Increasing the study's sample size would be the most straightforward way of increasing the statistical power. Alpha = 0.05 is the usual cut-off in a power calculation and it was used in this study. Increasing the effect size used in the power calculation would also increase the resulting power. However, the effect size should be the minimum value that would be of clinical importance. In this study, the effect size used in the power calculation was a between-group difference of 10%, which appears to be a reasonable number. Thus, increasing the study sample size (i.e., enrolling more patients) would be preferred here.

Question 6

a. SD provides the individual patient variability in responses around the mean value. Thus, the SEM in study A needs to be converted to SD using the equation SEM = SD/$\sqrt{n}$. The SD reported in study A would be: 5 = SD/$\sqrt{100}$. SD = 5 × 10 = 50. This SD is much higher than the SD reported in study B, so greater (wider) individual patient variability around the mean value was seen in study A.

Keep in mind that SEM should *not* be used in clinical studies in place of the SD. When used, always convert it to the SD.

b. This is the SD. Assuming the data are normally or near normally distributed, it indicates that ~68% of the individual patients' cholesterol concentrations in the study were within 185 − 18 mg% = **167 mg%** and 185 + 18 mg% = **203 mg%**. The remaining 32% of the patients' cholesterol concentrations in the study were outside of this range. The mean ±2 SD provides the range that contains ~95% of the individual study values, and the mean ±3 SD provides the range that contains ~99% of the individual patient values.

Question 7

b. The CI provides the range of values likely to contain the population value for that specified measure, at a certain level (%) confidence. In this example, the CI was reported for the difference in complete pain relief between toradol and naproxen (58% complete pain relief with toradol − 46% pain relief with naproxen = 12% difference). The CI is saying that in the population of patients outside the study (those persons who would still meet the specified inclusion/exclusion criteria), the difference between the two drugs might not be exactly 12%. But we are 95% confident that the difference could be anywhere from 7% *less* (i.e., it is a negative number) with toradol to up to 31% greater with toradol compared to naproxen.

Question 8

No. The CI contains the value of 0, which means that there could be no difference between complete pain relief with toradol and naproxen in the population. Since the CI contains this value of no difference, it is *not* statistically significant.

Question 9

a.

$$OR = \frac{\dfrac{24 \text{ (no. of patients with kidney disease with losartan)}}{323 \text{ (no. of patients with no kidney disease with losartan)}}}{\dfrac{40 \text{ (no. of patients with kidney disease with atenolol)}}{280 \text{ (no. of patients with no kidney disease with atenolol)}}}$$

$$= 0.074/0.143 = 0.52$$

b.

$$RR = \frac{\dfrac{24 \text{ (no. of patients with kidney disease with losartan)}}{347 \text{ (no. of patients taking losartan)}}}{\dfrac{40 \text{ (no. of patients with kidney disease with atenolol)}}{320 \text{ (no. of patients taking atenolol)}}}$$

$$= 0.069/0.125 = 0.55$$

c. NNT = 1 / ARR. The ARR with losartan is 12.5% − 6.9% = 5.6% or 0.056. Thus, NNT = 1 / 0.056 = 17.86 = 18 (*always* round up to nearest whole number).

This NNT means that 18 patients would need to be treated with losartan instead of atenolol for 12 months (the study duration) to prevent one additional case of kidney disease.

d. RRR = 1 − RR = 1 − 0.55 (calculated in part b) = 0.45 or 45%. Note how much higher the RRR appears compared to the ARR (5.6%) because the RRR is a proportion.

The other way to calculate the RRR is as follows:

$$RRR = \frac{\text{atenolol (\%)} - \text{losartan (\%)}}{\text{atenolol (\%)}}$$

$$RRR = 12.5\% - 6.9\% / 12.5\% = 0.45 = 45\%$$

Question 10

Intention-to-treat (intent-to-treat) was used in this study since data from all patients initially enrolled in the study, regardless of whether or not they completed the study, were included in the data analysis. This is generally the preferred data-handling method. Although not specified in this example, the study likely used last observation carried forward or a type of imputation calculation to include in their analyses the missing data points from the patients who dropped out.

When the intention-to-treat analysis is used with many drop-outs, the calculated efficacy is likely to be reduced. Why? Because if the drug was not taken for the full amount of time by the patients who dropped out, the drug might not have had the opportunity to show its full benefit. For studies such as this, it is useful for the investigators to report their results using both the intention-to-treat and exclusion of subjects approaches. This allows the reader to determine the effects, if any, of the data-handling method on the results seen.

Chapter 11

Question 1

The noninferiority margin should be the largest clinically acceptable difference in efficacy between treatment groups. If the study finds the efficacy difference to be smaller than the specified margin, it would indicate that the new treatment is at least as good as the active control (i.e., the new treatment is noninferior to the control).

Question 2

The null hypothesis for a noninferiority study is that the new treatment is inferior to the active control. No, it does not have the same meaning in a "superiority" study. The null hypothesis for a "superiority" study is that there is no difference between/among the treatments studied.

Question 3

b.

Question 4

CIs. They are generally reported for the efficacy results in a noninferiority study. In order for the new treatment to be noninferior to the control, the lower value of the treatment CI should not fall below the specified noninferiority margin (or the upper limit should not be above this margin, depending on which efficacy value is being subtracted from the other) in order for the new treatment to be considered noninferior to the control.

Chapter 12

Question 1

a, b, c, and e are correct. A meta-analysis is a type of systematic review that, in addition to being rigorous and structured, uses formal statistical techniques to quantitatively combine the results of separate (independent), but similar studies, into an overall, combined outcome measure. By doing this, a meta-analysis produces an outcome measure with a sample size beyond that of the individual studies involved. The result is an increased statistical power for detecting small but potentially significant differences. Meta-analyses are performed for several additional purposes, including allowing results to be applied to a greater variety of patients than a smaller, separate study might be able to do by itself, and answering questions that an individual study did not consider.

Question 2

A cumulative meta-analysis involves reperforming the meta-analysis each time a new study is identified to calculate a new, combined (pooled) outcome measure. A cumulative meta-analysis can graphically illustrate how an outcome measure becomes more precise (with a narrower confidence interval (CI) – refer to Chapter 9 for more information about CIs) as the number of patients increases. It can be used to identify statistically significant treatment effects sooner than might be required for a large individual study to be completed.

Question 3

This is incorrect. A meta-analysis is "observational" in nature in that it uses the findings from already completed studies to calculate a combined outcome measure statistically. Thus, it is subject to several possible limitations and biases. Consider that each study included in a meta-analysis might differ from the others in a number of important ways, such as study design; use of blinding; dosage, administration, or duration of therapy; the inclusion/exclusion criteria, or outcome measures used. The investigators conducting the meta-analysis might also fail to locate and include relevant studies, which could skew the results in a particular direction.

Question 4

Yes. Publication bias can occur when a meta-analysis only includes published studies; published studies can be more likely to show significant (positive; favorable) results than unpublished studies. Since Medline was used to locate relevant studies, the investigators limited their meta-analysis to only published studies. The investigators should have attempted to locate unpublished studies by searching for relevant information presented at conferences, consulting controlled trial registries that include published and unpublished sources, such as the Cochrane Central Register of Controlled Trials, one of the databases in the Cochrane Library (accessed at: http://www.thecochranelibrary.com/view/0/index.html), or by contacting other investigators in the subject area who might know if any unpublished findings exist.

Question 5

A sensitivity analysis is done to check for possible heterogeneity (important differences) or bias in a meta-analysis resulting from certain features of the studies included. A sensitivity analysis calculates a combined outcome measure with and without certain types of studies included, to determine if excluding studies significantly affects the outcome measure. In the example described, the sensitivity analysis might first calculate a pooled OR with small studies included, followed by calculating the OR again with small studies excluded, to determine the resulting effect on the OR. If there is no significant effect on the pooled OR by dropping the small studies, it means there was no apparent heterogeneity or bias based on study size. The sensitivity analysis can be repeated in a similar manner for other study design features to determine if any appear to cause significant heterogeneity or bias among the studies used in the meta-analysis.

Question 6

Yes, the National Guideline Clearinghouse (NGC) is an example of such a database. The NGC is maintained by the Agency for Healthcare Research and Quality, with the stated purpose "to provide physicians and other health professionals, health care providers, health plans, integrated delivery systems, purchasers, and others an accessible mechanism for obtaining objective, detailed information on clinical practice guidelines and to further their dissemination, implementation, and use." In order to be included in the NGC, a practice guideline must meet specific inclusion criteria. Refer to the NGC (www.ngc.gov) for detailed information about the database and how to use it to locate specific guidelines of interest. Other sources can also be used to locate clinical practice guidelines; these are found in Table 12.2.

Bibliography

Abate M A (2012). Clinical drug literature. In: Allen L V (ed.) *Remington: The Science and Practice of Pharmacy*, 22nd edn. Philadelphia, PA: Pharmaceutical Press, pp. 1877–1892.
Alpert J S (2010). Why are we ignoring guideline recommendations? *Am J Med* 123: 97–98.
American Society of Health-System Pharmacists (1996). ASHP guidelines on the provision of medication information by pharmacists. *Am J Health-Syst Pharm* 53: 1843–1845.
Balistreri W F (2007). Landmark, landmine, or landfill? The role of peer review in assessing manuscripts. *J Pediatr* 151: 107–108.
Blommel M L, Abate M A (2007). A rubric to assess critical literature evaluation skills. *Am J Pharm Educ* 71: article 63.
Bogner R H, Giovenale S (2012). Information resources in pharmacy and the pharmaceutical sciences. In: Allen L V (ed.) *Remington: The Science and Practice of Pharmacy*, 22nd edn. Philadelphia, PA: Pharmaceutical Press, pp. 37–46.
Bolton S, Hirsch R (2012). Statistics. In: Allen L V (ed.) *Remington: The Science and Practice of Pharmacy*, 22nd edn. Philadelphia, PA: Pharmaceutical Press, pp. 497–540.
Briggs G C, Freeman R K, Yaffe S J (2011). *Drugs in Pregnancy and Lactation*, 9th edn. Philadelphia, PA: Lipincott Williams and Wilkins.
Bruce N, Pope D, Stanistreet D (2008). *Quantitative Methods for Health Research*. West Sussex: John Wiley.
Chinchilli V M (2007). General principles for systematic reviews and meta-analyses and a critique of a recent systematic review of long-acting beta-agonists. *J Allergy Clin Immunol* 119: 303–306.
Davies H T O, Crombie I K (2009). What are confidence intervals and p-values? What is...? Hayward Group. http://www.whatisseries.co.uk/whatis/, (accessed October 15, 2012).
DePoy E, Gitlin L N (2011). *Introduction to Research. Understanding and Applying Multiple Strategies*. St. Louis, MO: Elsevier Mosby.
Egger M, Smith G D (1998). Bias in location and selection of studies. *BMJ* 316: 61–66.
Fischer J M (1980). Modificaton to the systematic approach to answering drug information requests. *Am J Hosp Pharm* 37: 470, 472–476.
Food and Drug Administration. Running clinical trials. http://www.fda.gov/ScienceResearch/SpecialTopics/RunningClinicalTrials/default.htm (accessed October 15, 2012).
Friedman L M, Furberg C D, DeMets D L (2010). *Fundamentals of Clinical Trials*. New York, NY: Springer.
Guyatt G, Rennie D, Meade M O, Cook D J, eds (2008). *Users' Guides to the Medical Literature. Essentials of Evidence-based Clinical Practice*. New York, NY: McGraw-Hill.
Hopkins W G. A new view of statistics. http://www.sportsci.org/resource/stats/errors.html (accessed October 17, 2012).
Institute of Medicine (1990). Clinical practice guidelines: directions for a new program. In: Field M J, Lohr K N (eds) Washington, DC: National Academy Press.
Institute of Medicine (2011). Standards for systematic reviews. Accessed at: http://www.iom.edu/Reports/2011/Clinical-Practice-Guidelines-We-Can-Trust/Standards.aspx, October, 2012.

Janosky J E (2005). Use of the single subject design for practice based primary care research. *Postgrad Med J* 81: 549–551.

Lang T A, Secic M (2006). *How to Report Statistics in Medicine*. Philadelphia, PA: American College of Physicians.

Leon A C (2011). Comparative effectiveness clinical trials in psychiatry: superiority, noninferiority, and the role of active comparators. *J Clin Psychiatry* 72: 1344–1349.

Lesaffre E (2008). Superiority, equivalence, and non-inferiority trials. *Bull NYU Hosp Joint Dis* 66: 150–154.

Medical Library Association. A user's guide to finding and evaluating health information on the web. http://www.mlanet.org/resources/userguide.html (accessed June 5, 2012).

MedlinePlus. MedlinePlus guide to healthy web surfing. http://www.nlm.nih.gov/medlineplus/healthywebsurfing.html (accessed October 17, 2012).

Moncur R A, Larmer J C (2009). Clinical applicability of intention-to-treat analyses. *Evidence Based Med* 6: 39–41.

Motulsky H (2010). *Intuitive Biostatistics*. New York, NY: Oxford University Press.

Piaggio G, Elbourne D R, Altman D G, Pocock S J, Evans S J W (2006). Reporting of noninferiority and equivalence randomized trials. *JAMA* 295: 1152–1160.

Robinson A, Kardos M, Kimball A B (2012). Physician Global Assessment (PGA) and Psoriasis Area and Severity Index (PASI): why do both? A systematic analysis of randomized controlled trials of biologic agents for moderate to severe plaque psoriasis. *J Am Acad Dermatol* 66: 369–375.

Schulz K F, Grimes D A (2002). Blinding in randomised trials: hiding who got what. *Lancet* 359: 696–700.

Taylor H (2010). "Cyberchondriacs" on the rise? The Harris poll #95, August 4, 2010. http://www.harrisinteractive.com/NewsRoom/HarrisPolls/tabid/447/mid/1508/articleId/448/ctl/ReadCustom%20Default/Default.aspx (accessed October 17, 2012).

US Department of Health and Human Services, Office for Human Research Protections. Institutional Review Board guidebook. http://www.hhs.gov/ohrp/archive/irb/irb_introduction.htm (accessed October 17, 2012).

US National Library of Medicine (2010). Databases, resources, and APIs. Accessed at: http://wwwcf2.nlm.nih.gov/nlm_eresources/eresources/search_database.cfm, October 17, 2012.

Vavken P (2011). Rationale for and methods of superiority, noninferiority, or equivalence designs in orthopaedic, controlled trials. *Clin Orthop Relat Res* 469: 2645–2653.

Vickers A (2010). *What is a P-Value Anyway?* Boston, MA: Addison-Wesley.

Vidal L, Shavit M, Fraser A, Paul M, Leibovici L (2005). Systematic comparison of four sources of drug information regarding adjustment of dose for renal function. *BMJ* 331: 263.

Vitry A L (2007). Comparative assessment of four drug interaction compendia. *Br J Clin Pharmacol* 63: 709–714.

Watanabe A S, McCart G, Shimomura S, Kayser S (1975). Systematic approach to drug information requests. *Am J Hosp Pharm* 32: 1282–1285.

Wright S G, LeCroy R L, Kendrach M G (1998). A review of three types of biomedical literature and the systematic approach to answer a drug information request. *J Pharm Pract* 11: 148–162.

Index

Note: Page references in fxx refer to Figures; those in txx refer to Tables

A Practical Guide to Contemporary Pharmacy Practice, 20
absolute risk increase (ARI), 128, 133, 134
absolute risk reduction (ARR), 105, 122, 126, 132
active controls, 74, 88, 143
adherence
　key points, 69, 81, 88, 89, 139
　worked examples, 83
administration routes, 80, 88, 138
adverse effects, 80, 81, 89, 139
adverse reactions, 18
　see also adverse effects
Agency for Health Care Research and Quality (AHRQ), 45
AHFS DI *see* American Society of Health-System Pharmacists
AHRQ *see* Agency for Health Care Research and Quality
AIDSinfo, 45
alpha (level of significance), 105, 117, 120, 132
　see also Type I errors
alternative hypotheses, 64, 65, 67, 92
alternative medicine resources, 21, 48
American Society of Health-System Pharmacists (AHFS DI), *AHFS Drug Information*, 14, 16, 18
analysis of variance (ANOVA), 93, 94, t94

ANOVA *see* analysis of variance
Applied Therapeutics: The Clinical Use of Drugs, 25
ARI *see* absolute risk increase
ARR *see* absolute risk reduction
Art, Science, and Technology of Pharmaceutical Compounding, 20
assignment to interventions, 77
authors
　clinical studies, 61
　critiquing published studies, 138
　Internet resources, 41, t43
　investigators, 61, 62, 66, 67
　potential conflicts of interest, 63, 64
averages, arithmetic, 110

background information
　question's to ask, 4
　systematic approach, 4
baseline/demographic comparisons, 3, 88, 140
before and after design, 74, 75, f75, t76
　see also time series design
bell-shaped curves (Gaussian distributions), 92, f92
beta (β) (level of significance), 105, 115, 118, t119, 132
　see also Type II errors
bias
　meta-analyses types, 154, 157
　potential conflicts of interest, 63

bias (*continued*)
   published studies, 63
   study methodology, 70, 72, 81
   tertiary resources, 8
blinding
   critiquing published studies, 139
   double-blinding, 78, 79, 88, 90
   peer review, use in, 62
   single-blinding, 78, 79, 89
   triple-blinding, 78, 88
   types, 69, 76, 77, 88–90
   unblinding, 79, 80, 89
   worked examples, 79
BNF see British National Formulary
Bonferroni statistical method, 94
Boolean operator, search strategies, 30
British National Formulary (BNF), *BNF for Children*, 23

CAM *see* complementary and alternative medicine
case-control study design, 54, 55, t56, 59
case reports, 53
case series, 53
categorical measurement, 86
Centers for Disease Control and Prevention (CDC), 45
central tendency, measures, 106, t108, 131
   critiquing published studies, 140
   mean, 106, t108, 110, 131
   median, 107, t108, 131, 140
   mode, 107, t108
   worked examples, 107
Chi-square test, 95
CI *see* confidence intervals
CINAHL, secondary resources, 33
citation, 29, 64, 154
classification, question handling, 6
Clinical Pharmacology, tertiary resources, 14
clinical practice guidelines
   databases, 156, t157, 159
   development, 151, 155
   locating, t157
clinical significance, 106, 121, 133, 140

ClinicalTrials.gov, 46
cluster sampling, 71
Cochrane Library, 33, 34
code of conduct, 42, 51
coefficient of determination (r^2), 99
cohort studies, 54, 55, t56, 58
   concurrent, 55
   historical design, 55
   nonconcurrent design, 55
   prospective design, 55
   retrospective design, 55
compatibility, parenteral resources, 19
compendia resources, 13
complementary and alternative medicine (CAM), 48
compliance *see* adherence
concurrent cohort studies *see* cohort studies
concurrent control, design, 74, 75, f75, t76, 77, 88
   critiquing published studies, 138
concurrent medication considerations, 81, 88, 89, 139
confidence intervals (CI)
   definition/interpretation, 105, 111, 113, 122, 132
   noninferiority studies, 146
   self-assessment questions, 135, 136
   statistical significance, 122
   worked examples, 114
conflicts of interest, 61–63, 66
   authors, 63, 64
   critiquing published studies, 138
   investigators, 62, 66, 67
confounding variables, 84
consumer information needs, 39
ConsumerLab, 49
contingency tables, 96
continuous measurements, 87
controlled experimental studies, *see also* experimental clinical studies
   advantages, t76
   control types, 69, 74, 88, 90
   critiquing published studies, 137
   disadvantages, t76
   explanatory studies, 53

controlled vocabulary searches,
    secondary resources, 32
controls
    critiquing published studies, 138
    equivalence studies, 143
    noninferiority studies, 143
    types in experimental studies, 69,
        74, 76, 88, 90
convenience sampling see sampling,
    types
copyright issues, t43, 51
correlation coefficients, 98, 101, f98
correlation, statistical analysis, 91, 98,
    f98 101, 139
cross-over
    critiquing published studies, 138
    study design, 74, 75, f75, t76,
        77, 88
cross-sectional studies, 54, 57, t57
cumulative meta-analysis, 152
cut-off see probability values

data handling, drop-outs, 105, 128,
        133, 134
    critiquing published studies, 140
    errors, 106
    exclusion of subjects/per protocol
        method, 129, 130, 134
    intent-to-treat/intention-to-treat,
        129, 130, 134
    worked examples, 130
Database of Reviews of Effectiveness
    (DARE), 33
databases
    bias in meta-analysis, 154
    clinical practice guidelines, 156,
        t157, 159
    Internet resources, 48, 49
    natural medicines, 49
    practice guidelines, 156, t157,
        159
    secondary resources, 33, 37
    tertiary resources, 17, 18, 24
    toxicity, 48
demographic/baseline comparisons, 3,
    88, 140
dependent variables, 69, 83

descriptive clinical studies, 53
    case reports, 53
    case series, 53
Dietary Supplements Labels Database,
    46
Directory of Information Resources
    Online (DIRLINE), 46
discussion section, 105, 131, 133, 134,
    140
dispersion, variability measures, 108
Doctor's Guide to Medical and Other
    News, 48
dosage, treatment considerations, 23,
    80, 88
    blinding, 78
    critiquing published studies, 138
double-blinding, 78, 79, 88, 90
double-dummy, 78
drop-outs see data handling, drop-outs
drug concentrations, treatment
    considerations, 80
Drug Facts and Comparisons, 15
Drug Information Handbook, 15
Drug Interaction Facts, 19
drug interactions, 19
*Drug Interactions Analysis and
    Management*, 19
Drugdex database, Micromedex, 17,
    18
Drugs@FDA database, 46
Drugs in Pregnancy and Lactation, 24

editorial boards, 61, 66, 138
effect size, 118, 119, 130
efficacy
    noninferiority studies, 145
electronic resources
    secondary resources, 31, 34, 35
    tertiary resources
        AHFS Drug Information, 14
        Drug Facts and Comparisons, 15
        Drug Information Handbook, 16
        Micromedex, 17
        Natural Standard, 21
eligibility criteria, 69, 70, 87, 88, 138
Embase, 34
Emerging Infectious Diseases, 45

English-language bias in meta-analyses, 154
enrollment considerations, 71, 87, 138
Epocrates, 50
equivalence studies, 143, 144
   controls, 143
   examples, 144
   interpretation, 143
errors
   see also type I errors; type II errors
   noninferiority studies, 147
   self-assessment questions, 135
   worked examples, 120
Evaluations of Drug Interactions, 20
exclusion criteria, 70, 87, 88, 138
exclusion of subjects/per protocol data handling, 129, 130, 134
experimental clinical studies, 53, 59, 71, 89
   see also controlled experimental studies
explanatory clinical studies, 53
   see also experimental clinical studies; observational clinical studies
extemporaneous compounding resources, 20
Extemporaneous Formulations for Pediatric, Geriatric, and Special Needs Patients, 20
external validity, 85, 89

FDA *see* Food and Drug Administration
Fisher's exact test, 95, 96
Food and Drug Administration (FDA), 17, 35, 46, 72
Friedman tests, 96
funnel plots in meta-analysis, 154

Gaussian distributions, 92, f92
Goodman and Gilman's: The Pharmacological Basis of Therapeutics, 25
Google Scholar, 31, 34
government-sponsored Internet resources, 45

Handbook on Injectable Drugs, 19
Harriet Lane Handbook, 23
Hawthorne effect, 85
Health Hotlines, 47
Health On the Net Foundation, 42, 51
Healthfinder, 47
herbal medicine resources, 21
heterogeneity, meta-analysis, 153
historical cohort studies *see* cohort studies
historical controls *see* controls, types in experimental studies
HONcode, Internet resources, 42, 51
hypotheses, 92
   alternative hypothesis, 64, 65, 67, 92
   clinical study results and interpretation, 105, 114
   noninferiority studies, 147
   null hypothesis, 64–66, 115, 147
   one-tailed (one-sided) alternative hypothesis, 65, 92
   probability values, 115, t119, 132
   published studies, 61, 64–67
      critiquing studies, 138
   two-tailed (two-sided) alternative hypothesis, 65

IDIS *see* Iowa Drug Information System
inclusion criteria, 69, 87, 88, 138
independent variables, 69, 83
Index Nominum: International Drug Directory, 22
indexing services, secondary resources, 29
inference, clinical studies, 105, 111
information-specific tertiary resources, 18
informed consent, 69, 72, 88
injectable drug resources, 19
Institutional Review Boards (IRB), 72, 88
intent-to-treat/intention-to-treat, data handling, 129, 130, 134
internal validity, 85, 89
International Classification of Diseases, 31, 35

international drug product resources, 22, 27
International Pharmaceutical Abstracts (IPA), 31, 34
Internet resources
　Agency for Health Care Research and Quality, 45
　AHFS Drug Information, 16
　AIDSinfo, 45
　alternative medicine, 48
　author considerations, 41, t43
　Centers for Disease Control and Prevention, 45
　characteristic considerations, 40
　ClinicalTrials.gov, 46
　complementary and alternative medicine, 48
　consumer information needs, 39
　ConsumerLab, 49
　content validity, 41, t43
　copyright, 51, t43
　current information considerations, 41, t43
　databases
　　natural medicines, 49
　　toxicity, 48
　design, 42, t44
　Dietary Supplements Labels Database, 46
　Directory of Information Resources Online, 46
　Doctor's Guide to Medical and Other News, 48
　Drug Information Handbook, 16
　Drugs@FDA database, 46
　Emerging Infectious Diseases, 45
　Epocrates, 50
　evaluation, 40
　Food and Drug Administration, 46
　government-sponsored sites, 45
　Health Hotlines, 47
　health information, evaluation, 40
　Healthfinder, 47
　HONcode, 42, 51
　information needs, 39
　lactation, 48
　LactMed, 48
　MD Consult, 49
　Medical Library Association, 42
　medication information, 40
　MedlinePlus, 42, 47
　Medscape, 49
　mobile devices, 50
　Morbidity and Mortality Weekly Reports, 45
　National Cancer Institute, 48
　National Center for Complementary and Alternative Medicine, 48
　Natural Medicines Comprehensive Database, 49
　Natural Standard, 49
　News agency sites, 50
　nongovernment sites, 48
　Orange Book, 46
　organization, 42, t44
　pregnancy, 48
　Preventing Chronic Diseases, 45
　privacy policy, 42, t44
　product-testing sites, 49
　quality, 39, 40
　question handling, 42, t43
　reference considerations, 42, t44
　reliability, 39
　RxList, 49
　search strategies, 39
　self-assessment questions, 51
　source considerations, 41, t43
　tertiary resources, 16
　Toxnet, 48
　travel information, t43, 51
　USP Verified, 49
　validity, 41, t43
　veterinary products, 47
　worked examples, 40
interquartile range, 105, 108, 109, 132, 133
intervention assignment, 77
investigators *see* authors
Iowa Drug Information System (IDIS), 31, 35
IPA *see* International Pharmaceutical Abstracts
IRB *see* Institutional Review Boards

journals, 63
  critiquing published studies, 138
  editorial boards, 61, 66
  peer review, 62, 66

King Guide to Parenteral Admixtures, 19
Kruskal–Wallis test, 96

Lact-Med, 24, 48
lactation and pregnancy resources, 24, 26, 48
language bias, in meta-analysis, 154
limits, search strategies, 32
Litt's Drug Eruptions and Reactions Manual, 18
Loansome Doc program, 30
location, study considerations, 3
logistic regression, 100

Mann–Whitney U/Wilcoxon rank-sum test, 96
Martindale: The Complete Drug Reference, 16, 22
masking *see* blinding
McNemar tests, 95, 96
MD Consult, 49
mean, 106, t108, 110, 131, 133
  see also central tendency, measures
  self-assessment questions, 134
measurement
  categorical measurement, 86
  central tendency, 106, t108, 131
  characteristics, 69, 84, 88
  continuous measurements, 87
  nominal measurement levels, 86, 87
  ordinal measurements, 86, 87
  worked examples, 87
median, 107, t108, 131, 133, 134, 140
  see also central tendency, measures
Medical Library Association, 42
Medical Subject Headings (MeSH), 31, 34, 36
Medications and Mother's Milk, 24
Medline, 29, 31, 34, 36
MedlinePlus, 42, 47
Medscape, 49

MeSH *see* Medical Subject Headings
meta-analyses
  biases, 154, 157
  locating, 151, t155
  preparation, 151, 152
  problems, 151, 153, 158
  purpose, 151, 152
  resources, 151
  self-assessment questions, 158
  worked examples, 157
Meyler's Side Effects of Drugs, 18
Micromedex, 17, 18, 25
missing data, handling, 105
MMWR *see* Morbidity and Mortality Weekly Reports
mobile devices, 50
mode, 107, t108, 134
  see also central tendency, measures
Morbidity and Mortality Weekly Reports (MMWR), 45
multiple comparison tests, 94
multiple regression, 100

n-of-1 experimental studies, 54
National Cancer Institute (NCI), 48
National Center for Complementary and Alternative Medicine (NCCAM), 48
National Library of Medicine's MedlinePlus, 42, 47
Natural Medicines Comprehensive Database, 21, 49
Natural Standard, 21, 49
NCCAM *see* National Center for Complementary and Alternative Medicine
NCI *see* National Cancer Institute
neonatal dosing resources, 23
News agency sites, 50
nominal measurements, 86, 87
nonconcurrent cohort studies *see* cohort studies
noncontrolled experimental studies, 53
nongovernment Internet resources, 48
noninferiority studies, 143, 145
  controls, 143
  efficacy, 145

hypothesis, 147
implementation, 149
interpretation, 143
margins thresholds, 146
null hypothesis, 147
performance reasons, 145
probability values, 146, 147
problems, 143, 148
self-assessment questions, 150
type I errors, 147
type II errors, 147
worked examples, 146
nonparametric tests, 93, 95, t97, 99, 100
   Chi-square test, 95
   Fisher's exact test, 95, 96
   Friedman tests, 96
   Kruskal–Wallis test, 96
   Mann–Whitney U/Wilcoxon rank-sum test, 96
   Wilcoxon signed-rank test, 96
   worked examples, 96, 97
normal (Gaussian) distributions, 92, f92
null hypothesis, 64–66, 115, 147
number needed to harm (NNH), 106, 122, 128
number needed to treat (NNT), 105, 122, 126, 128, 133
   critiquing published studies, 140
   self-assessment questions, 136

objectives
   published study evaluations, 61, 64–66
      critiquing, 138, 139
      worked examples, 65
observational clinical studies, 53, 54, t56, 59,
   see also case-control studies; cohort studies; cross-sectional studies
odds ratio (OR), 122, 124, 127, 132, 134
   self-assessment questions, 136
one-tailed (one-sided) hypothesis, 64, 92
one-way statistical measures, 94, t94

online resources see Internet resources
open-labeled/nonblinded studies, 78, 89
OR see odds ratio
Orange Book, 46
ordinal measurements, 86, 87
organization of Internet resources, 42, t44
outcomes, 83, 88–90
   primary outcomes, 83, 88, 139
   reliability, 84, 85, 89
   secondary outcomes, 83, 88, 139
   sensitivity, 84
   specificity, 84
   validity, 84, 85, 89
   worked examples, 84
outliers, 106

P values see probability values
paired data, 93, 94, 97
parallel control design, 74, 75, f75, t76, 77, 88
parametric tests, 92, 93, 97, t97, 100
   worked examples, 95
parenteral product compatability/stability, 19
patient drop-outs see data handling, drop-outs
patient-specific background information, 4, 6
PDR see Physician's Desk Reference
Pearson r/Pearson product-moment r, 99
Pediatric & Neonatal Dosage Handbook, 23
Pediatric Drug Formulations, 21
pediatric resources, 20, 23
peer review, 61, 62, 66, 67
Pharmacotherapy Principles and Practice, 25
Pharmacotherapy: A Pathophysiological Approach, 25
Physician's Desk Reference (PDR), 18
placebo controls, 74, 76, 79
plural terms, secondary resources, 32
population of interest considerations, 70, 71, 87
population values, 112, 132

post hoc tests, 94
power
   clinical study methodology, 73, 74, 88
   clinical study results, 115, 118, 130
   critiquing published studies, 140
   effect size, 130
   noninferiority studies, 149
   sample size, 73, 132
   self-assessment questions, 135
   worked examples, 74
practice guidelines
   databases, 156, t157, 159
   development, 151, 155
   locating, 151, t157
   purpose, 155
   resources, 151
pregnancy and lactation resources, 24, 26, 48
Preventing Chronic Diseases, 45
primary literature, 9, 13, 29
primary outcomes, 83, 88, 139
probability (P) values, 133
   cut-off, 116
   hypothesis testing, 115, t119, 132
   noninferiority studies, 146, 147
   self-assessment questions, 135
   statistical significance, 115, 116, t119, 121
   worked examples, 116
product-testing information, 49
prospective cohort studies, 55
publication bias in meta-analysis, 154

quality of Internet resources, 39, 40
question handling
   background information, 4
   classification, information needs, 6
   demographics, 3
   evidence-based approaches, 1, t3
   implementation, 10
   information needs, 1, t3
   Internet resources, 42, t43
   patient-specific background information, 4, 6
   response provision, 9
   restating questions, 9
   systematic approach, 1, 2, t3
   tertiary resources, 13
   worked examples, 5

r see coefficient of determination
random assignment, 69, 71, 77, 138
random sampling, 71, 87
range, 105, 108
reference considerations, 42, t44
regression analysis, 91, 100
relative risk (RR), 105, 122–124, 127, 134
relative risk reduction (RRR), 105, 122, 125, 132, 134, 136
reliability
   Internet resources, 39
   outcome measures, 84, 85, 89
Remington: The Science and Practice of Pharmacy, 24
repeated measures, ANOVA, 94, t94
Reprorisk database, *Micromedex*, 17, 25
requester demographic issues, 3
resources, *see also* Internet resources; secondary resources; tertiary resources
retrospective case-control studies, 55
retrospective cohort studies, 55
risk, 105, 122, 123, 132, 134
   self-assessment questions, 136
   worked examples, 127
risk reduction, 105, 122, 134
RR *see* relative risk
RRR *see* relative risk reduction
RxList, Internet resources, 49

sample size
   clinical study methodology, 73, 74, 88
   noninferiority studies, 148
   power, 73, 132
   type II errors, 118, 119
sampling, 69, 71, 87
   cluster sampling, 71
   random sampling, 71, 87
   stratified random sampling, 71
   systematic sampling, 71

sd *see* standard deviation
search strategies
    Boolean operators, 30
    evaluations, 8
    information needs, 7, 13, 29, 39
    Internet resources, 39
    limits, 32
    resources, 7
    review articles, 8
    secondary resources, 7, 8, 29
    synthesizing skills, 8, 9
    systematic approaches, 7
    tertiary resources, 7, 13
secondary outcomes, 83, 88, 139
secondary resources
    Boolean operators, 30
    CINAHL, 33
    citations, 29
    Cochrane Library, 33, 34
    controlled trials, 33, 34
    controlled vocabulary, 32
    databases, 37
        CINAHL, 33
        Cochrane Library, 33
    disadvantages, 29, 32
    electronic versions, 31, 34, 35
    Embase, 34
    evaluation, 8
    examples, 33
    full-text articles, 29
    Google Scholar, 31, 34
    identifying search terms, 30
    indexing services, 29
    information needs, 7, 8, 29
    International Classification of Diseases, 31, 35
    International Pharmaceutical Abstracts, 31, 34
    Iowa Drug Information System, 31, 35
    limits, 32
    Loansome Doc program, 30
    Medical Subject Headings, 31, 34, 36
    Medline, 29, 31, 34, 36
    plural terms, 32
    search strategies, 7, 8, 29
    self-assessment questions, 37
    synonyms, 32
    terminology, 30
    truncation, 32
    worked examples, 31
selection bias
    clinical study methodology, 70, 72
    clinical study types, 55, t56, 60
    meta-analysis, 154
self-assessment questions
    clinical study results, interpretation and conclusions, 134
    clinical study types, 59
    experimental study types, 59
    information needs, 11
    Internet resources, 51
    meta-analysis, 158
    noninferiority studies, 150
    observational study types, 59
    published studies, 66
    secondary resources, 37
    statistical analysis, 101
    study methods, 89
    systematic evidence approaches, 11
    tertiary resources, 26
sem *see* standard error of the mean
sensationalism, Internet resources, 40, 41
sensitivity, outcome measures, 84
side effects, 19
simple random sampling, 71
simple regression, 100
single-blinding, 78, 79, 89
single-subject research design, 54
size considerations *see* sample size
skewed data, 107, 140
Spearman rank-order, 99
specificity, outcome measures, 84
stability
    parenteral products, 19
standard deviation (sd), 108, 120, 132, 133
    worked examples, 111
standard deviation (sd), 64
standard error of the mean (sem), 64, 110, 132, 133
    worked examples, 111

statistical inference, 105, 111
statistical power *see* power
statistical significance
    clinical significance, 106, 121, 133, 140
    confidence intervals, 122
    critiquing published studies, 140
    cut-off, 116
    probability values, 115, 116, t119, 121
Stockley's Drug Interactions, 20
stratified random sampling, 71
Student's *t*-test, 93, 94
synonyms, in searching secondary resources, 32
systematic evidence/information handling, 1, 2, t3, 5, 7, 11
    background information, 4
    patient-specific background information, 4, 6
    question handling, 2
    responses, 3, 9
systematic reviews
    implementation, 158
    preparation, 151
    purpose, 151
    resources, 151
    source identification, 151, t155
    worked examples, 156
systematic sampling, 71

t-test, statistical analysis, 93, 94
tertiary resources
    A Practical Guide to Contemporary Pharmacy Practice, 20
    adverse reactions, 18
    AHFS Drug Information, 14, 16, 18
    alternative medicines, 21
    Applied Therapeutics: The Clinical Use of Drugs, 25
    Art, Science, and Technology of Pharmaceutical Compounding, 20
    bias, 8
    BNF for Children, 23
    Clinical Pharmacology, 14
    compatibility of parenteral products, 19
    compendia, 13
    databases
        Micromedex, 17, 18
        pregnancy and lactation, 24
    dosing, 23
    Drug Facts and Comparisons, 15
    Drug Information Handbook, 15
    Drug Interaction Facts, 19
    drug interactions, 19
    Drug Interactions Analysis and Management, 19
    Drugdex database, Micromedex, 17, 18
    Drugs in Pregnancy and Lactation, 24
    electronic versions
        AHFS Drug Information, 14
        Drug Facts and Comparisons, 15
        Drug Information Handbook, 16
        Micromedex, 17
        Natural Standard, 21
    evaluation, 7, 8
    Evaluations of Drug Interactions, 20
    extemporaneous compounding, 20
    Extemporaneous Formulations for Pediatric, Geriatric, and Special Needs Patients, 20
    FDA labeling, 17
    geriatric patients, 20
    Goodman and Gilman's: The Pharmacological Basis of Therapeutics, 25
    Handbook on Injectable Drugs, 19
    Harriet Lane Handbook, 23
    herbal medicines, 21
    Index Nominum: International Drug Directory, 22
    information-specific resources, 18
    injectables, 19
    international drug products, 22, 27
    Internet
        AHFS Drug Information, 16
        Drug Information Handbook, 16
    King Guide to Parenteral Admixtures, 19

Index | 195

Lact-Med, 24
lactation, 24
Litt's Drug Eruptions and Reactions Manual, 18
Martindale: The Complete Drug Reference, 16, 22
Medications and Mother's Milk, 24
Meyler's Side Effects of Drugs, 18
Micromedex, 17, 18, 25
natural medicines, 21
Natural Medicines Comprehensive Database, 21
Natural Standard, 21
neonatal dosing, 23
parenteral products
  compatibility, 19
  stability, 19
Pediatric Drug Formulations, 21
pediatric patients, 20, 23
Pharmacotherapy Principles and Practice, 25
Pharmacotherapy: A Pathophysiological Approach, 25
Physician's Desk Reference, 18
pregnancy, 24
primary literature, 13
Remington: The Science and Practice of Pharmacy, 24
Reprorisk database, Micromedex, 17, 25
search strategies, 7, 13
self-assessment questions, 26
side effects, 19
special needs patients, 20
stability, parenteral products, 19
Stockley's Drug Interactions, 20
The Review of Natural Products, 22
therapeutics, 25
Toxnet database, 24
Trissel's Stability of Compounded Formulations, 21
The Pharmacological Basis of Therapeutics, 25
The Review of Natural Products, 22
therapy-related questions, 1
thresholds, noninferiority studies, 146

time series design, 74, 75, f75, t76, 88
toxicity, 48
Toxnet database, 24, 48
travel information, t43, 51
treatment considerations
  adherence, 69, 81
  administration routes, 80
  adverse effects, 80, 81
  compliance *see* adherence
  concentration, 80
  concurrent, medications, 81
  critiquing published studies, 138
  dosage, 80
  drug concentrations, 80
  therapy duration, 80
triple-blind studies, 78, 88
Trissel's Stability of Compounded Formulations, 21
truncation, secondary resources, 32
two-tailed (two-sided) alternative hypothesis, 65
two-way measures, statistical analysis, 94, t94
type I errors, 105, 115, **117**, t119, 120, 147
type II errors, 105, 115, 118, t119, 147
types of clinical studies, 53

unblinding, 79, 80, 89
unpaired data, 94, 97
USP Verified, 49

validity
  critiquing published studies, 140
  information needs, 7–9
  Internet resources, 41, t43
  outcome measures, 84, 85, 89
  primary literature, 9
  secondary resources, 8
  tertiary resources, 7, 8
variability measures
  clinical study results, 105, 108, 132
  critiquing published studies, 140
  dispersion, 108
  interquartile range, 105, 108, 109, 132
  range, clinical studies, 105, 108

variability measures (*continued*)
    self-assessment questions, 134, 136
    spread, 108, 132
    standard deviation, 64, 108, 120,
        132, 133
    standard error of the mean, 64, 110,
        132, 133
    type II errors, 120
    worked examples, 109, 111
variables
    clinical study methodology, 69, 83
    correlation, 98
    critiquing published studies, 140
    regression, 100
variance in clinical studies, 109
variance, square root of, 109
verification considerations, 5
veterinary, Internet resources, 47

websites *see* Internet resources
width of confidence intervals, 113
Wilcoxon rank-sum test, 96
Wilcoxon signed-rank test, 96
worked examples
    adherence (compliance), 83
    blinding, 79
    central tendency of data, 107
    confidence intervals, 114
    control types, 76

data handling, 130
eligibility criteria, 70
hypothesis, 65
information needs, 5
Internet resources, 40
measurement scales/levels, 87
meta-analysis, 157
noninferiority studies, 146
nonparametric tests, 96, 97
number needed to treat, 128
objectives, 65
observational clinical studies, 58
odds ratio, 127
outcome measures, 84
parametric tests, 95
patient drop-outs, 130
power, 74
probability values, 116
question handling, 5
relative risk, 127
risk, 127
sample size, 74
sampling, 71
secondary resources, 31
standard deviation, 111
standard error of the mean, 111
statistical power, 74
systematic reviews, 156
variability measures, 109, 111